AF368531

Advanced CMOS Integrated Circuit Design and Application

Advanced CMOS Integrated Circuit Design and Application

Editors

Jong-Ryul Yang
Seong-Tae Han

MDPI • Basel • Beijing • Wuhan • Barcelona • Belgrade • Manchester • Tokyo • Cluj • Tianjin

Editors
Jong-Ryul Yang
Department of Electronic
Engineering, Yeungnam
University
Korea

Seong-Tae Han
Electrophysics Research Center,
Korea Electrotechnology
Research Institute
Korea

Editorial Office
MDPI
St. Alban-Anlage 66
4052 Basel, Switzerland

This is a reprint of articles from the Special Issue published online in the open access journal *Sensors* (ISSN 1424-8220) (available at: https://www.mdpi.com/journal/sensors/special_issues/desig_appl).

For citation purposes, cite each article independently as indicated on the article page online and as indicated below:

LastName, A.A.; LastName, B.B.; LastName, C.C. Article Title. *Journal Name* **Year**, *Volume Number*, Page Range.

ISBN 978-3-0365-3477-0 (Hbk)
ISBN 978-3-0365-3478-7 (PDF)

Contents

About the Editors

Jong-Ryul Yang received his B.S. degrees in electrical engineering and material science from Ajou University, Suwon, South Korea, in 2003 and his Ph.D. degree in electrical engineering from the Korea Advanced Institute of Science and Technology, Daejeon, South Korea, in 2009. From February 2009 to October 2011, he worked as a senior engineer in the System LSI Division, Samsung Electronics, Yongin, South Korea. From November 2011 to August 2016, he was a senior researcher in the Advanced Medical Device Research Division, Korea Electrotechnology Research Institute, Ansan, South Korea. From March 2012 to August 2016, he was an associate professor in the Department of Energy and Power Conversion Engineering, University of Science and Technology, Ansan, South Korea. Since September 2016, he has been an associate professor in the Department of Electronic Engineering, Yeungnam University, Gyeongsan, South Korea. He is a senior member of IEEE, a lifetime member of KIEES, KIEE, and IEIE, and a board member of MDPI *Sensors*. His research interests include RF/millimeter-wave/terahertz-wave circuits and systems, specifically for CMOS detector ICs, high-power millimeter-wave transmitters, sub-terahertz imaging systems, and miniaturized radar sensors.

Seong-Tae Han received his B.S. degree in physics education in 1999, and both his M.S. and the Ph.D. degrees in physics, in 2001 and 2005, respectively, from Seoul National University (SNU), Seoul, Korea. As a research assistant, he participated in developing ultrawide-band TWTs for electronic warfare and satellite communication at X-band. He demonstrated the first working MEMS-fabricated vacuum electronic device (a folded waveguide TWT) at Ka-band. Dr. Han was a researcher with the Research Institute of Basic Science, SNU, in 2005, where his research focused on novel vacuum electron devices employing recent innovations, such as photonic crystals and cold cathodes based on nano/MEMS technologies. In September 2005, he joined in the Plasma Science and Fusion Center, Massachusetts Institute of Technology (MIT), Cambridge, MA, as a postdoctoral research associate, where he worked on sub-THz gyrotrons for DNP/NMR research and ECH transmission lines for ITER. Since 2008, he has been a senior and principal researcher with the Korea Electrotechnology Research Institute, Korea (KERI), where he leads programs for developing high-power and high-frequency gyrotrons for active denial systems and real-time nondestructive inspection systems, respectively. His research interests cover high-power microwaves and charged particle beams, as well as their industrial applications. Currently, he serves as the director of Electrophysic Research Center with KERI and a professor with the Department of Energy and Power Conversion Engineering in the University of Science and Technology.

Review

Switched-Biasing Techniques for CMOS Voltage-Controlled Oscillator

Cheol-Woo Kang [1], Hyunwon Moon [2,*] and Jong-Ryul Yang [1,*]

[1] Department of Electronic Engineering, Yeungnam University, Gyeongsan, Gyeongbuk 38541, Korea; cwkang@yu.ac.kr
[2] Department of Electronic Engineering, Daegu University, Gyeongsan, Gyeongbuk 38453, Korea
* Correspondence: mhw@daegu.ac.kr (H.M.); jryang@yu.ac.kr (J.-R.Y.); Tel.: +82-53-850-6651 (H.M.); +82-53-810-2495 (J.-R.Y.)

Abstract: A voltage-controlled oscillator (VCO) is a key component to generate high-speed clock of mixed-mode circuits and local oscillation signals of the frequency conversion in wired and wireless application systems. In particular, the recent evolution of new high-speed wireless systems in the millimeter-wave frequency band calls for the implementation of the VCO with high oscillation frequency and low close-in phase noise. The effect of the flicker noise on the phase noise of the VCO should be minimized because the flicker noise dramatically increases as the deep-submicron complementary metal-oxide-semiconductor (CMOS) process is scaled down, and the flicker corner frequency also increases, up to several MHz, in the up-to-date CMOS process. The flicker noise induced by the current source is a major factor affecting the phase noise of the VCO. Switched-biasing techniques have been proposed to minimize the effect of the flicker noise at the output of the VCO with biasing AC-coupled signals at the current source of the VCO. Reviewing the advantages and disadvantages reported in the previous studies, it is analyzed which topology to implement the switched-biasing technique is advantageous for improving the performance of the CMOS VCOs.

Keywords: CMOS; voltage-controlled oscillator; switched-biasing; flicker noise; phase noise; current source; figure-of-merit

Citation: Kang, C.-W.; Moon, H.; Yang, J.-R. Switched-Biasing Techniques for CMOS Voltage-Controlled Oscillator. *Sensors* **2021**, *21*, 316. https://doi.org/10.3390/s21010316

Received: 19 November 2020
Accepted: 2 January 2021
Published: 5 January 2021

Publisher's Note: MDPI stays neutral with regard to jurisdictional claims in published maps and institutional affiliations.

1. Introduction

A voltage-controlled oscillator (VCO) is a key component in a frequency synthesizer that generates local oscillator (LO) signals for frequency conversion in a radio-frequency (RF) transceiver [1–3]. A VCO-based readout circuit, which is that the output voltage of the sensing core is applied to the node of the VCO tuning voltage, has merit to achieve a low sensitivity level and a high signal-to-noise ratio compared to the amplifier-based readout circuit [4,5]. In addition, high integration and low power consumption of the VCO-based readout circuit are advantageous for implementing a large-scale sensor array [6,7]. Radar sensors that monitor the change of electromagnetic-wave between the transmitted and received signals generated from the VCO remotely measure the distance, velocity, and vital-signs in real time [8–10]. Various sensors using VCOs require low phase noise characteristics in VCOs [11–13]. The frequency synthesizer such as the phase-locked loop (PLL) is conventionally used to reduce the phase noise, but the noise characteristics still remain at the output signal of the frequency synthesizer because the loop bandwidth of the PLL in the synthesizer is generally determined to be between 100 and 500 kHz [14,15].

A complementary metal-oxide-semiconductor (CMOS) process is a standard fabrication technology to implement electrical circuits; it can integrate control, logic, analog, and RF circuits into a single-chip system [16]. Moreover, the recent CMOS device designed in the several nanometer-scale shows competitive performances in transconductance (g_m) and minimum noise figure (NF_{min}), compared to the compound semiconductor device [17]. However, the transistor device implemented in the up-to-date CMOS process exhibits an increase in the flicker noise, which intrinsically depends on the physical structure of the channel and the flicker corner frequency, where the magnitudes of the flicker and white noises present equal increases up to several MHz or more [18,19]. The flicker noise is called "1/f noise" because the noise is increased as the frequency in the channel decreases [20]. The nanometer-scaled CMOS technology has advantages such as high integration, low power consumption, and high operating frequency, but it has the disadvantage of noise deterioration in the low-frequency band owing to the increase in the flicker noise [14].

The reduction in the flicker noise is a major issue in VCO design using the CMOS process because the phase noise of the VCO is mainly determined by the noise in the low-frequency region [21]. Many studies have been conducted to improve the reduction in the CMOS VCO phase noise performance caused by the flicker noise. Biasing techniques for core transistors have been widely used to prevent the degradation of the VCO phase noise by the flicker noise of the oscillator core transistors [22,23]. A VCO using core transistors biased at class C operation is a representative technique that reduces the contribution of the flicker noise effect from the core transistor at the output [24–26]. A resonant filter at the second harmonic also minimizes the noise effect from the core transistors, but the large chip size and the tuning range limit cause other issues in the design of the VCO using this technique [27]. The increase in the chip area can be reduced by implementing the filter using the common-mode resonance of the LC tank in a cross-coupled LC oscillator, although the issue of the tuning range limit cannot be solved [15,28,29]. The performance degradation by the noise of the core transistors is reduced by these techniques, but the contribution of the flicker noise caused by the current source, which is used to constantly supply the DC bias current in the core transistors, remains in the output characteristics of the VCO. The flicker noise by the current source dramatically increases the close-in phase noise of the VCO owing to the nonlinear characteristics of the VCO [30]. A simple method to reduce the effect of the flicker noise from the transistors constituting the current source is to design the VCO using only voltage biasing, that is, without using any current sources [14,31]. However, the core current of the VCO can easily deviate from the designed value depending on the power supply variations when current biasing is not used. It can also be observed that the oscillator becomes more sensitive to ground noise [21]. A switched-biasing technique has been proposed to reduce the flicker noise effect of the current source based on the periodic behavior of the differential VCO [32,33]. The up-conversion behavior of the flicker noise of the current source can be fundamentally eliminated using the switched-biasing technique, which is based on the periodic operation of the differential VCO [34].

In this paper, we review operations, features, and implementation examples of the VCO using the switched-biasing technique as a method to reduce the flicker noise effect generated by the current source. The effect of improvement in the phase noise by this technique is examined based on a comparison of the VCO performance obtained by the topologies implementing switched-biasing. In particular, it is analyzed whether the switched-biasing technique is useful for generating the millimeter-wave signals, which are increasingly used in various applications. Section 2 introduces the first proposal of a switched-biasing technique for the reduction of the flicker noise effect of CMOS transistors and describes the advantages of the technique in the VCO design. In Section 3, the switched-biasing technique is classified into three topologies based on how the biasing voltage of the current source is configured, and the features and implementation examples of VCOs using each topology are presented. Section 4 discusses the advantages and disadvantages of each topology by comparing the CMOS VCOs with switched-biasing techniques. Based on the discussion results, the applicability of the switched-biasing technique for the millimeter-wave signal generation, which is mandatory to the high-resolution radar sensors operating at 20 GHz or more, is examined.

2. Switched-Biasing Technique

As the deep-submicron CMOS process is scaled down, the low-frequency noise (especially the flicker noise) of the MOSFET becomes more important in the design of CMOS RF transceivers. It has long been known that the flicker noise is generated in a variety of homogeneous semiconductor bulks and is observed in various devices, such as a vacuum tube, diode, and MOSFET [35,36]. Various research works have been conducted to identify the cause of the flicker noise and to clearly understand its characteristics clearly [32,33,35–43]. To predict the flicker noise phenomenon generated in MOSFETs, Hooge published a carrier mobility fluctuation (CMF) model, in which the flicker noise is caused by the mobility fluctuation of free carriers in the device [37]. McWhorter suggested a carrier number fluctuation (CNF) model, where the low frequency noise of the MOSFET is generated by the fluctuation in the number of charge carriers in the device [38]. The two presented models were useful for understanding the physical mechanism of the flicker noise, but their limitation is that they can only be applied to the long-channel devices. The flicker noise in short-channel devices is mainly considered to be due to the random telegraph signal (RTS) noise generated by the $Si-SiO_2$ interface because as the size of the devices is scaled down, the device operation is predominantly represented by the movement of each charge carrier [39,41,44].

Research on reducing the intrinsic flicker noise of MOSFETs began in the early 1990s. Bloom and Nemirovsky first suggested that the flicker noise of the MOSFET could be reduced by cycling between inversion and accumulation of the device [40]. They explained that the device noise in the on-state can be reduced when the off-state exists before the on-state. Dierickx and Simoen revealed that the flicker noise reduction by inversion-to-accumulation cycling is related to the emptying of traps at the interface that generates RTS noise [41]. Based on the principle of inversion-to-accumulation cycling, Gierkink et al. proposed a switched-biasing technique [32]. Figure 1 shows the operating principle of the switched-biasing technique [33]. The "operational state" in Figure 1 means that the MOSFET operates at the inversion state, to facilitate the flow of current between the drain and the source. The drain–source current does not flow at the "rest-state" of the MOSFET because the bias voltage at the gate is lower than the threshold voltage of the device. A reduction in the flicker noise can be expected by the periodic operation between the two states of the device and is verified with a simple mathematical analysis. Assuming a duty cycle of 50%, the drain–source current of the MOSFET by the switching operation can be expressed as the multiplication of the flicker noise and a square-wave signal $m(t)$ with the duty cycle,

$$m(t) = \frac{1}{2} + \frac{2}{\pi}\sin(\omega_{sw}t) + \frac{2}{3\pi}\sin(3\omega_{sw}t) + \frac{2}{5\pi}\sin(5\omega_{sw}t) + \cdots, \tag{1}$$

where ω_{sw} is the angular frequency of the switching operation. Because the power spectral density (PSD) of the noise in the low-frequency band is determined by the convolution of the DC component of $m(t)$ and the flicker noise, the switched-biasing technique can decrease the PSD by 6 dB compared to the constant-biasing technique. In addition, several studies have confirmed that the flicker noise is further reduced when the transistor is sufficiently turned off (i.e., deep accumulation). This reduction is known to be caused by the elimination of the long-term-memory effect associated with the flicker noise [32,33,44]. The analysis of the operating characteristics and principle shows that the PSD due to low-frequency noise depends on the bias state at the time and the bias history in a periodic operation [44]. This phenomenon was verified in both NMOS and PMOS because the carrier type does not affect the operating principle [43].

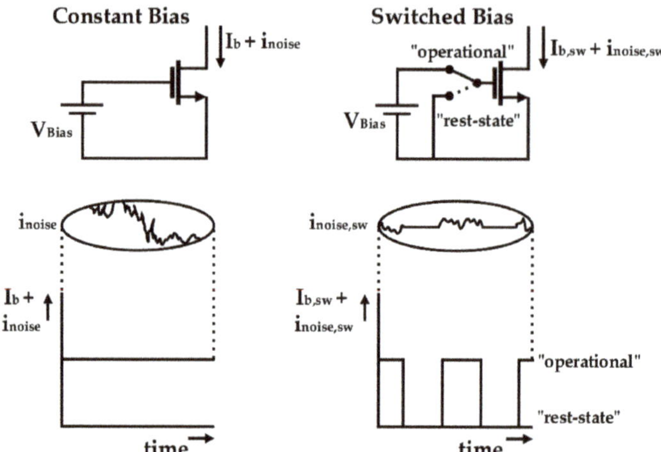

Figure 1. Conceptual diagram of switched-biasing technique (reproduced with permission from the author, reducing MOSFET $1/f$ noise and power consumption by switched biasing; published by IEEE, 2000) [33].

Flicker noise reduction by using the switched-biasing technique has been applied to various circuit designs, such as amplifiers and frequency mixers requiring low-noise characteristics [32–34,45,46]. In particular, the switched-biasing technique can be useful for the oscillator to improve the performance because the output signal in the oscillator is generated by the periodic switching operation of the transistor. After Gierkink et al. showed that the effect of flicker noise in the ring oscillator can be reduced by applying the switched-biasing technique to the DC bias control, Kluperink et al. demonstrated improvement in closed-in phase noise by switching the current source in the sawtooth oscillator [33]. Boon et al. showed that the switched-biasing technique can improve the phase noise of the LC oscillator by reducing the up-conversion of the flicker noise of the current source [34]. As shown in Figure 2, the switched-biasing of the current sources is implemented by the output oscillation signals, and the DC level of the current sources is self-biased by the structural characteristics of a CMOS LC-VCO. The VCO shown in Figure 2 shows phase noise improvement of 6 dB and 3 dB at 10 kHz offset compared to the VCO with the fixed-biasing current source and the VCO without the current source, respectively [34].

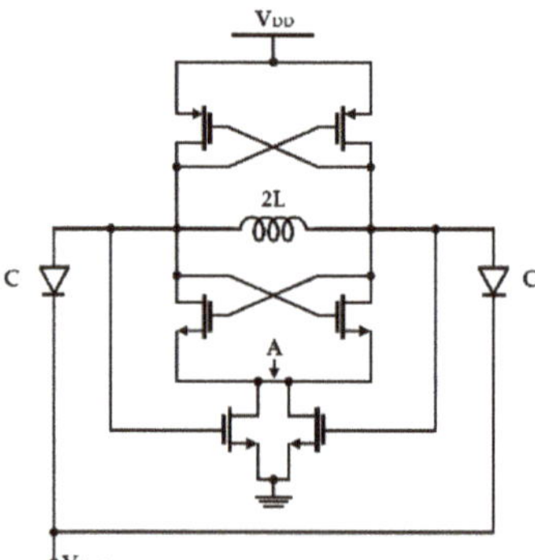

Figure 2. Schematic of a CMOS LC-VCO with the switched-biasing technique to the current source (reproduced with permission from the author, RF CMOS low-phase-noise LC oscillator through memory reduction tail transistor; published by IEEE, 2004) [34].

The characteristics that the current source modulated by the switched-biasing technique is effective in improving the phase noise of a VCO was proved using a theoretical analysis based on a mathematical model [47]. The proposed theoretical analysis is based on the impulse sensitivity function (ISF) theory, which can explain the phase noise contribution depending on the output voltage swing of the VCO [48]. The proposed analysis in Figure 3 shows that the phase noise of a VCO can be greatly improved by additionally injecting the bias current to the VCO core transistors at the time when the voltage swing of the VCO is maximized or minimized. This phenomenon is based on the fact that the time when the output voltage of the VCO becomes the maximum or minimum has the minimum sensitivity to the phase shift [47]. Figure 3b shows that the phase noise of the VCO can be minimized by the modulation signals of $2f_0$ in the current source compared to the fixed-biasing current source. It is caused that the bias currents of the cross-coupled transistors in the VCO using the switched-biasing current source are limited at a time of high phase-shift sensitivity and supplied at a time of low phase-shift sensitivity. The currents I_{d1} and I_{d2} supplied from the switched biasing current source are not supplied at the highly sensitive time in the phase noise where the output voltages $V_{o,n}$ and $V_{o,p}$ are crossed. Based on physical and theoretical interpretations, it can be verified that the switched-biasing technique improves the phase noise of the VCO by modulating the current source.

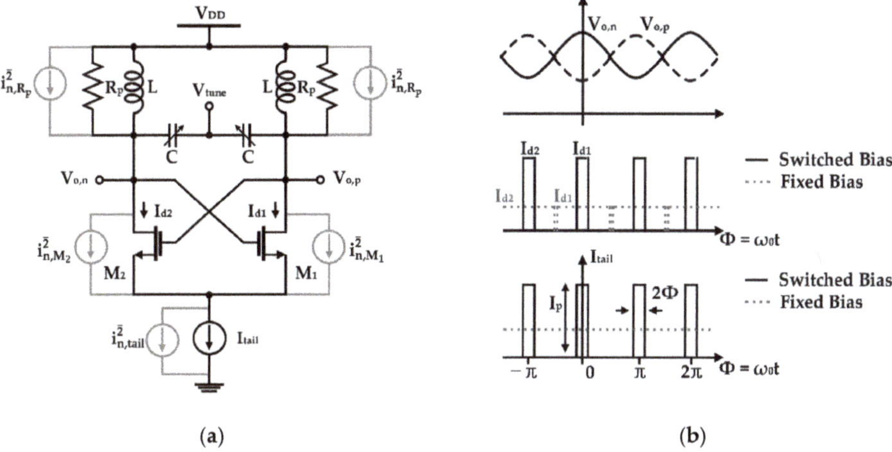

(a)   (b)

Figure 3. Analysis of VCO characteristics depending on the pulse modulation of the current source using the impulse sensitivity function theory: (**a**) schematic of the differential LC-VCO; (**b**) conceptual waveforms of the output voltages and drain currents of the VCO and the bias current by the pulse-modulated current source (reproduced with permission from the author, Tail current-shaping to improve phase noise in LC voltage-controlled oscillators; published by IEEE, 2006) [47].

3. Circuit Implementation

There are various design approaches to implement the switched-biasing technique for current source modulation in the VCO. Three design specifications in the current source should be considered in the implementation of the VCO with this technique:

- Modulation frequency and amplitude
- Modulation waveform
- DC bias voltage

The oscillation frequency of the VCO or the specific frequency generated from an external signal source can be used as the modulation frequency, which is related to the amount of flicker noise reduction [44,49,50]. The modulation amplitude should be sufficiently large to ensure periodic inversion-to-accumulation operations at the current source [51,52]. The modulation waveform affects whether the current source operates as hard-switching or soft-switching and the efficiency of the noise reduction [47]. A DC bias voltage should be determined as the specific value (e.g., the threshold voltage of the transistor constituting the current source) to obtain the effective switching operation, considering the modulation amplitude [52–54]. Based on these specifications, the proposed VCO design methods using the switched-biasing technique are divided into three topologies. Depending on whether the individual source for bias modulation is used or not, they are largely divided into external-biasing and self-biasing topologies. The self-biasing topology is further subdivided according to the usage of fixed or adaptive DC bias voltages. The detailed implementation methods of the switched-biasing technique applied to various VCO architectures are shown within the classification of these three topologies.

3.1. External-Biasing Topology

To verify the effectiveness of the flicker noise reduction phenomenon, in the initial study, the gate bias of the MOS device was externally applied [32,40,41]. Similarly, the switched-biasing technique can be implemented by externally controlling the gate bias of the current source of the VCO. In addition, the efficiency of the flicker noise reduction by the switched-biasing technique can be significantly improved in the VCO because the external signal generator can be optimally designed with the modulation frequency, amplitude, waveform including the duty cycle, and DC bias voltage.

Yoshida et al. proposed a structure that digitally controls the current source of a ring-type VCO using a switched-bias circuit (SBC), which is depicted in Figure 4 [50]. As shown in Figure 4b, the SBC is composed of two-level shifters and two identical bias branches (BC1 and BC2) and is configured to operate alternately according to the clock (CK) signal applied from the outside. In the VCO core shown in Figure 4a, current sources divided into three bits are placed on each side of the delay cell, and their gate bias is switched to a modulated signal (V_{cp} and V_{cn}) formed through the SBC. In addition, all current sources are biased to perform a triode operation because the probability of trap−detrap is less than that of the saturation operation and less flicker noise is generated [55]. The clock frequency of the SBC was set to 10 MHz. The purpose is to suppress spurious occurrences at the switching frequency despite the simulation result that noise reduction below 100 kHz is independent of the clock frequency. Effectively utilizing the SBC, the ring VCO improves the noise performance by 3 dB at 100 kHz [50].

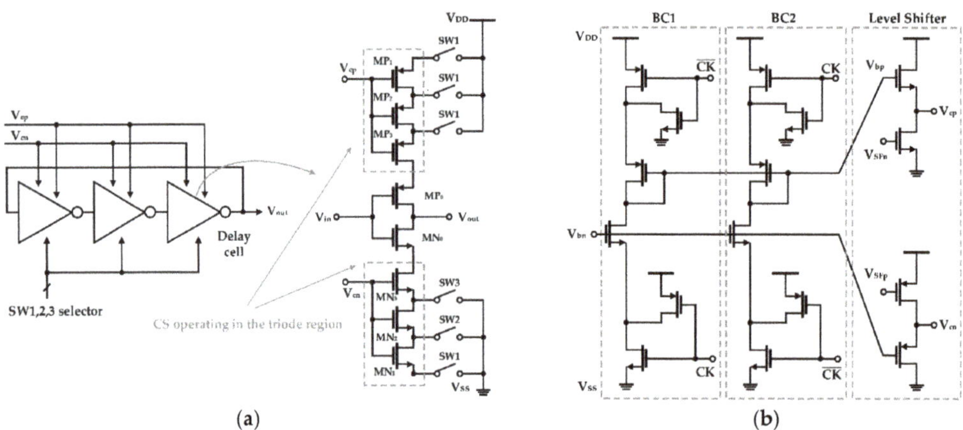

Figure 4. Ring VCO based on a switched-bias circuit (SBC): (**a**) block diagram of the ring VCO and schematic of the delay cell in the VCO; (**b**) schematic of SBC for current source modulation (reproduced with permission from the author, Low-voltage, low-phase-noise ring voltage-controlled oscillator using 1/f-noise reduction techniques; published by The Japan Society of Applied Physics, 2007) [50].

3.2. Self-Biasing Topology with the Fixed DC Voltage

A fully on-chip-type circuit cannot be configured by externally applying the modulation signal, and the design of the switched-bias technique is complicated because of the spurious dependence of the modulation frequency [50,56]. Although the external-biasing topology can supply the optimal modulation signal to the current source, the modulation signal correlated with the oscillation signal cannot be guaranteed to have an additional current injection at the time with the minimum sensitivity to the phase shift, as shown in Figure 3b. The phase noise of the oscillator can be further reduced by driving the current source modulated with the frequency which is the same as the oscillation frequency [34,47].

Jeong and Yoo proposed a method of applying switched-biasing to the current sources of each g_m stage and the coupled input stage in a CMOS quadrature-VCO (QVCO), as presented in Figure 5 [57]. When applied to a conventional current source coupled QVCO, the switched-biasing technique can be applied to the current source shared by the coupled input stage and the g_m stage. However, when the g_m stage and the coupled input stage share a current source, the oscillation waveform and the common source node waveform are misaligned because of the resistance in the transistor triode region and parasitic capacitances in the common source node (shown as Vs in Figure 5a). In this design, the source nodes of the cross-coupled pair and the coupled-input pair are separated for optimal alignment, and the QVCO waveform is shown in Figure 5b. As a result, the oscillation amplitude increased by 0.3 V (peak-to-peak) compared to the structure that shared the current source. Compared to the constant bias current and shared-current source method, phase noise improvements of 17 dB and 10 dB were shown in the simulation, respectively. In the measurement results, an improvement in the performance of 10 dB by the switched-biasing technique was verified, compared to that of the shared-current source method [57].

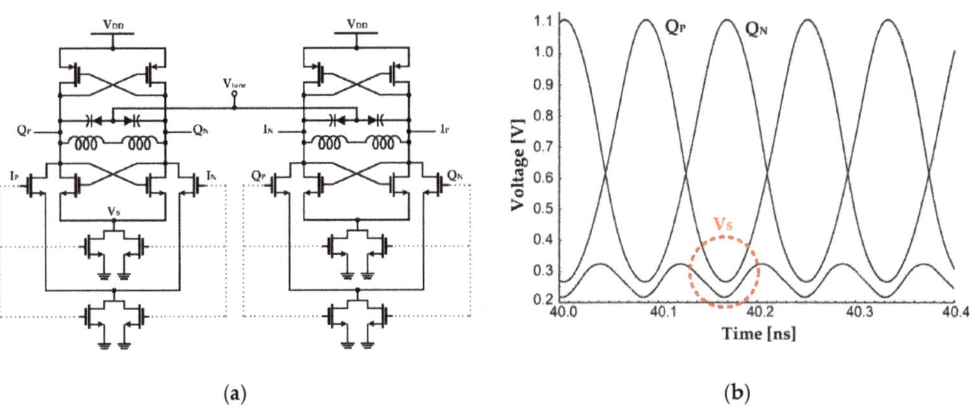

(a)   (b)

Figure 5. CMOS quadrature-VCO (QVCO) with separate current sources based on the switched-biasing technique: (a) schematic of the QVCO; (b) simulation waveforms at the outputs of the quadrature channel and the common node (reproduced with permission from the author, Low-phase-noise CMOS quadrature VCO; published by IEEE, 2006) [57].

Musa applied switched-biasing to the current source of a VCO operating near millimeter-wave, as shown in Figure 6 [51]. Unlike the CMOS structure, as it is an NMOS-only structure, an additional bias path (as shown in V_{Bias} of Figure 6a) for setting an appropriate DC bias level and a capacitor (as shown in C_F of Figure 6a) for coupling with the oscillation node are added. As depicted in Figure 6b, based on the ISF theory, the phase noise was improved through optimal current injection (i.e., the zero crossing point of the ISF) [47]. In addition, as the size of the feedback capacitance C_F determines the modulation signal amplitude, the capacitance was determined as an optimal value considering the trade-off between phase noise improvement and power consumption [51].

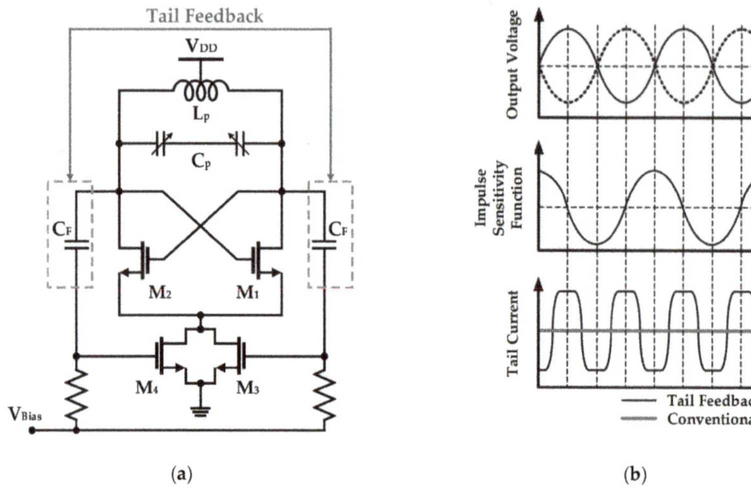

(a)

(b)

Figure 6. Millimeter-wave NMOS VCO using the switched-biasing technique: (**a**) schematic of the VCO; (**b**) waveform analysis of the VCO based on the ISF theory (reproduced with permission from the author, a low phase noise quadrature injection locked frequency synthesizer for mm-wave applications; published by IEEE, 2011) [51].

Huang and Kim proposed a self-biasing QVCO using the current source splitting (CSS) method, as illustrated in Figure 7a [52]. Unlike the conventional method of sharing a current source, it is designed to separate and deliver current to each cross-coupled transistor. This method has the advantage of being able to ignore noise caused by parasitic capacitance appearing at the common source node of a cross-coupled pair through separation of the corresponding node. In addition, the current source (NM_{5-6} in Figure 7a) and the cross-coupled pair (NM_{1-2} in Figure 7a) act as two cascode cross-coupled pairs, creating an effective negative resistance. As mentioned in Section 2 and as shown in Figure 7b, the long-term memory effect was eliminated by maximizing the modulation amplitude (i.e., VCO in the voltage-limited region), thus increasing the flicker noise reduction effect of the current source. To prove this, the result of the circuit simulation with which the flicker noise factor of the MOSFET was removed as compared to the measurement result of the fabricated QVCO, and similar phase noise improvement was confirmed [52].

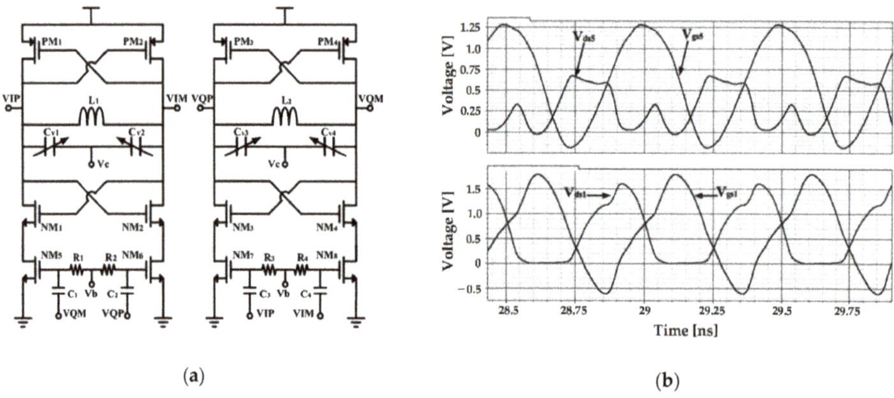

(a)

(b)

Figure 7. CMOS QVCO with split current sources: (**a**) schematic of the QVCO; (**b**) simulated waveforms at the drain–source and gate-source voltages of NM5 (top) and NM1 (bottom) (reproduced with permission from the author, Low phase noise self-switched biasing CMOS LC Quadrature VCO; published by IEEE, 2009) [52].

Chen et al. proposed a method to suppress the flicker noise generated from cross-coupled pairs by adding a source degeneration capacitor, as depicted in Figure 8a [58]. The degeneration capacitor (C_D in Figure 8a) is set to have a low impedance at the fundamental frequency and high impedance at a low frequency (i.e., flicker noise), as presented in Figure 8b [59]. In addition, using a filtering capacitor (as shown in C_f of Figure 8a), a low pass filter was constructed to remove noise from the bias path. In the simulation results, the closed-in phase noise of the VCO was improved by 2 dB using the current source modulation, but the structure using the degeneration capacitor improved 4, 11, and 7.5 dB at 10 kHz, 100 kHz, and 1 MHz, respectively [58].

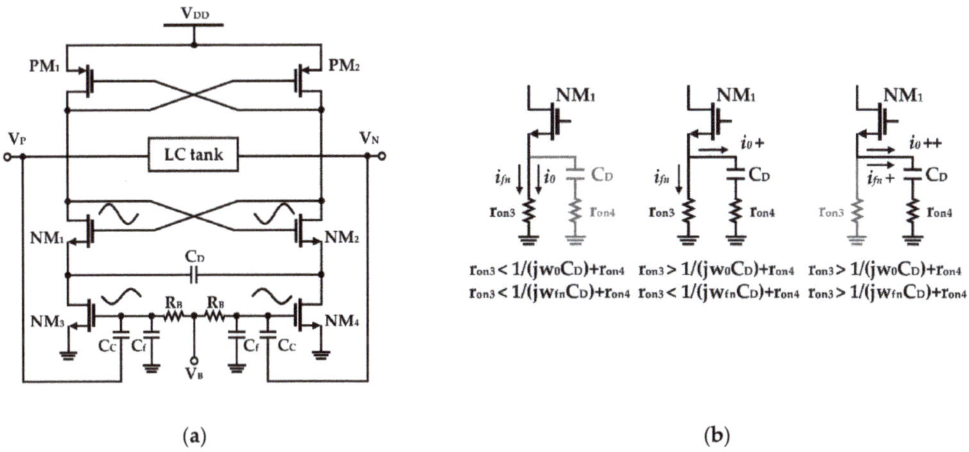

(a)   (b)

Figure 8. CMOS VCO including degeneration and filter capacitors: (**a**) schematic of the VCO; (**b**) current flows depending on the capacitance of CD—the usage of the optimal capacitance can divide the current into two paths, as shown in the middle case (reproduced with permission from the author, Reduction of $1/f^3$ phase noise in LC oscillator with improved self-switched biasing; published by Springer, 2015) [58].

Hsieh and Lin proposed adding a passive network between the current source and the VCO core to suppress the up-conversion of the second harmonic noise, as shown in Figure 9a [53]. As the second harmonic current of the common source node of the cross-coupled pair is up-converted and acts as noise, C_1, shown in Figure 9a, is connected in parallel with the current source to filter the second harmonic thermal noise [27]. Moreover, as the quality-factor (Q-factor) of the LC tank decreases during the period when the cross-coupled pair transistor operates in the triode, L_1 is added to increase the impedance of the common source node. As shown in Figure 9b, it was verified that the closed-in phase noise (100 kHz$-$1 MHz) characteristic improved to 3 dB when the passive network was added based on the same switched-biasing technique [53].

Based on the ISF theory, Mostajeran et al. proposed the ISF manipulation technique to reduce the flicker noise contribution by reducing the effective ISF of the tail current source, as shown in Figure 10 [60]. Considering that the ISF in the current source, it is necessary to implement two turn-offs during one oscillation period, and a separated current source structure was adopted in a similar manner to previous studies. By deep triode operation of the PMOS transistor as a current source, a low impedance path to the ground is formed so that less noise generated from the tail flows into the tank. Owing to switched biasing and the operation of the triode region of the PMOS, it can be observed that the effective ISF of the current source is reduced compared to the conventional NMOS current source, as depicted in Figure 10b. The measurement indicated an improvement of 17 dB in phase noise at 10 kHz and 8.2 dB at 1 MHz compared to phase noise with the structure using the conventional NMOS current source, and a very low flicker corner frequency of 10 kHz was confirmed [60].

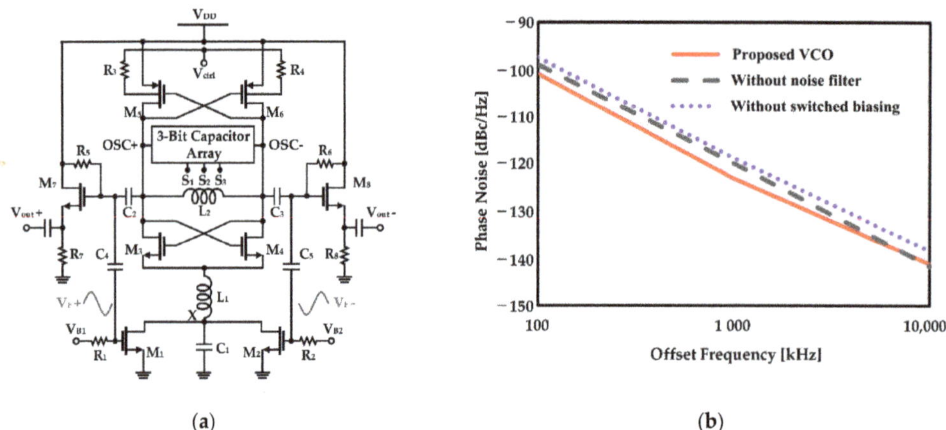

Figure 9. CMOS VCO with improved phase noise characteristics by adding a passive network: (**a**) schematic of the VCO; (**b**) phase noise of the VCO with the current source modulation and the noise filter (displayed as "proposed VCO"), the VCO using only the current source modulation (displayed as "without noise filter"), and the VCO using the constant bias voltage without the filter (displayed as "without switched biasing") (reproduced with permission from the author, A 0.7-mW LC voltage-controlled oscillator leveraging switched biasing technique for low phase noise; published by IEEE, 2019) [53].

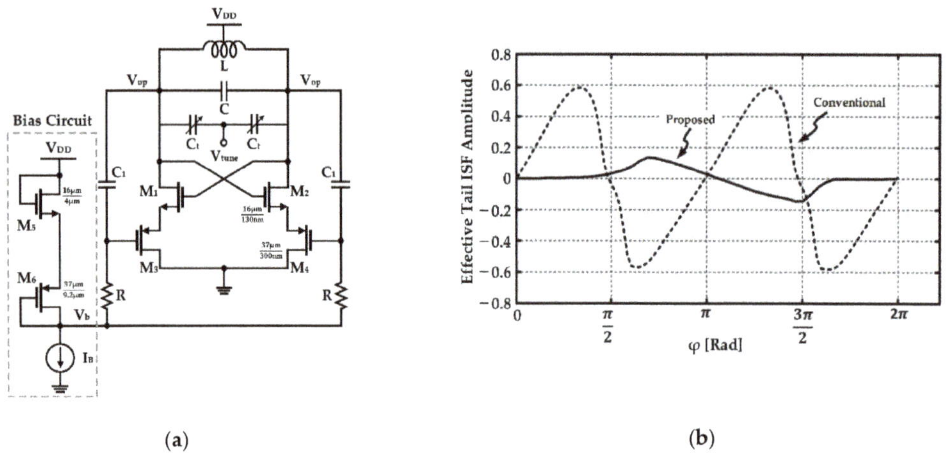

Figure 10. CMOS VCO based on the ISF manipulation technique: (**a**) schematic of the VCO including current sources using PMOS transistors; (**b**) waveforms of the effective tail ISF amplitude for the conventional current source using NMOS and the proposed current source using PMOS (reproduced with permission from the author, a 2.4 GHz VCO with FOM of 190 dBc/Hz at 10 kHz-to-2 MHz offset frequencies in 0.13 μm CMOS using an ISF manipulation technique; published by IEEE, 2015) [60].

Shasidharan et al. proposed a structure in which the switched-biasing technique is applied to a Class-C CMOS VCO, as shown in Figure 11 [54]. In this design, a source degeneration capacitor to suppress flicker noise of a cross-coupled pair and an auxiliary $-g_m$ stage to compensate for insufficient negative g_m were constructed. A current source using PMOS transistors was used to make a low impedance path, and an effective switching operation was achieved by the biasing at the sub-threshold voltage. Moreover, by adjusting the size of the current source appropriately, the parasitic capacitance C_{in} was designed to be an even-mode harmonic filter of the common node (V_{CM1-2} in Figure 11b). As it was designed to have a narrow conduction angle of 0.31π through simulation, the phase noise characteristic shows an improvement of 14 dB in the performance at 1 MHz offset compared to the case where the switched-biasing technique was not applied [54].

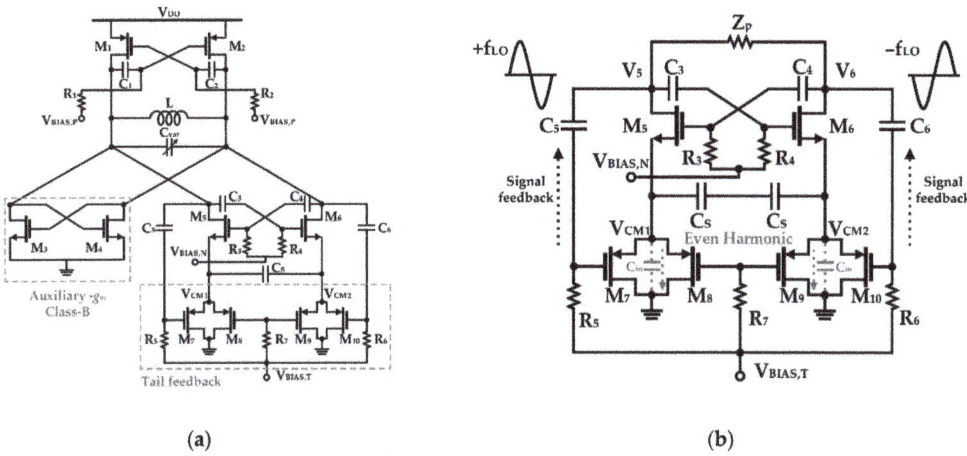

(a)   (b)

Figure 11. Class-C CMOS VCO using the switched-biasing technique: (**a**) schematic of the VCO; (**b**) bias conduction at the sub-threshold operation of the split current sources implemented by PMOS transistors (reproduced with permission from the author, A 2.2 to 2.9 GHz complementary class-C VCO with PMOS tail-current-source feedback achieving—120 dBc/Hz phase noise at 1 MHz offset; published by IEEE, 2019) [54].

Lee and Im applied the switched-biasing technique in a simple inverter delay cell based ring oscillator, which is depicted in Figure 12 [61]. As shown in Figure 12b, by self-biasing the current source of the CMOS inverter, the slope of the output waveform increases, and, as a result, the flicker noise is reduced and the oscillation swing is improved [62]. Compared to the topology without self-biasing, an improvement of 7 dB at 100 kHz and 11.5 dB at 1 MHz was verified through simulation [61].

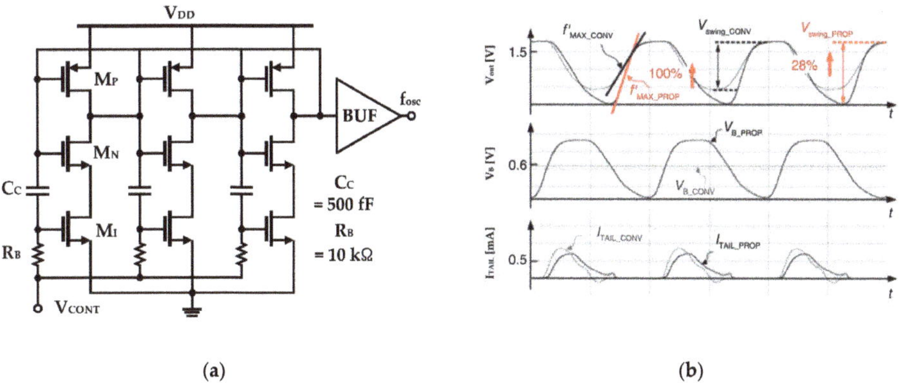

(a)   (b)

Figure 12. Ring VCO using the switched-biasing technique: (**a**) schematic; (**b**) simulated waveforms of the VCO output (top), bias voltage of the current source (middle), and tail current of the inverter cell (bottom) (reproduced with permission from the author, Low phase noise ring VCO employing input-coupled dynamic current source; published by IET, 2020) [61].

As the self-bias topology is coupled from the oscillation node, there is the advantage that an external modulation signal is not required. In addition, because it is implemented in the design of the VCO alone, the noise design can be controlled differently from the case of external bias. However, in some studies, the DC bias level is set near the threshold voltage of the current source for a proper switching effect [53,54,63]. Unsurprisingly, switching of the current source occurs after oscillation has begun, and, hence, an excessively low DC bias may not provide adequate starting conditions.

3.3. Self-Biased Topology with the Adaptive DC Voltage

To solve the start-up issue of the general self-bias topology, a method of adaptively adjusting the DC bias level of the current source according to the oscillation amplitude can be used. As mentioned in Section 2, because the current efficiency can be improved by reducing the conduction angle, a narrow conduction angle can be implemented by lowering the DC bias level as it approaches the steady state.

Min et al. suggested that the DC bias level of the current source can be adaptively adjusted according to the VCO oscillation amplitude by adding an auxiliary peak detector to the existing self-biasing topology, as shown in Figure 13 [63]. In this design, referring to Figure 13b, a separate cross-coupled pair (depicted in M_5-M_6 of Figure 13a) detects the negative peak of the oscillation waveform and induces charging to the capacitor (as shown in C_1 in Figure 13a). The charged voltage (V_k) changes the DC bias level of the current source, solving the start-up issue of the oscillator, and simultaneously reducing the oscillator amplitude variability, mainly due to the Q-factor change of the capacitor bank within the tuning range and the process, voltage, and temperature (PVT) variation of the circuit. The phase noise using the switched-biasing technique was improved to approximately 6 dB at 100 kHz in the simulation, compared to that using the constant-biasing technique [63].

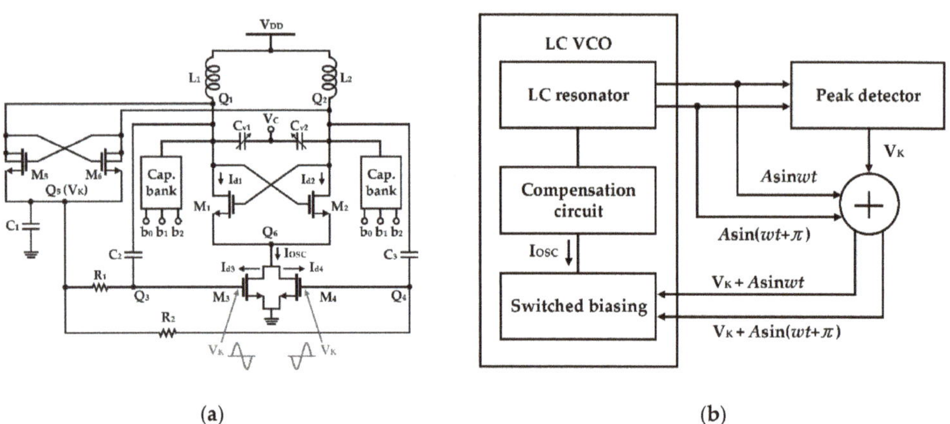

(a)   (b)

Figure 13. VCO using self-biasing with the adaptive DC voltage: (**a**) schematic; (**b**) operation of the self-biasing with the adaptive DC voltage from peak tracking at the oscillator outputs (reproduced with permission from the author, Low voltage CMOS LC VCO with switched self-biasing; published by Wiley, 2009) [63].

Narayanan and Okada presented the VCO architecture using a synchronized pulse generator to inject pulse-shaped waveforms into each current source, as shown in Figure 14 [64]. Unlike the self-biasing with the fixed DC voltage in which a modulation signal is injected directly through a capacitor, a rail-to-rail waveform is implemented using a two-stage inverter to narrow the conduction angle. As an additional method to reduce the conduction angle, an envelope tracking scheme called conduction angle control, shown in Figure 14a, was used. It lowers the DC bias level applied to the current source as the VCO oscillates, as indicated in Figure 14b, so that the conduction angle decreases as the VCO reaches a steady state. As the DC bias is approximated to the supply voltage when the current source supplies current, the triode operation is possible; thus, the noise generation of the current source is reduced compared to the case of the saturation operating point. Because of this design method, this VCO lowered the flicker noise corner to 700 Hz, and thus obtained the result of having a flattened figure-of-merit (FoM) in the range of 1 kHz−10 MHz. However, the intrinsic delay of the pulse generator clarifies the limitations of this method, as shown in Figure 14c. For proper current bias based on the ISF theory, a positive peak voltage must be delivered to the current source at the point where the ISF is zero crossing, but an indispensable mismatch occurs because of the corresponding delay. The result is shown in detail in Figure 14d. By adjusting the delay of the pulse waveform modeled with Verilog-A, when the delay exceeds $8/\pi$, the phase noise degradation occurs rapidly, and even when it reaches $4/\pi$, it can be observed that oscillation does not occur [64].

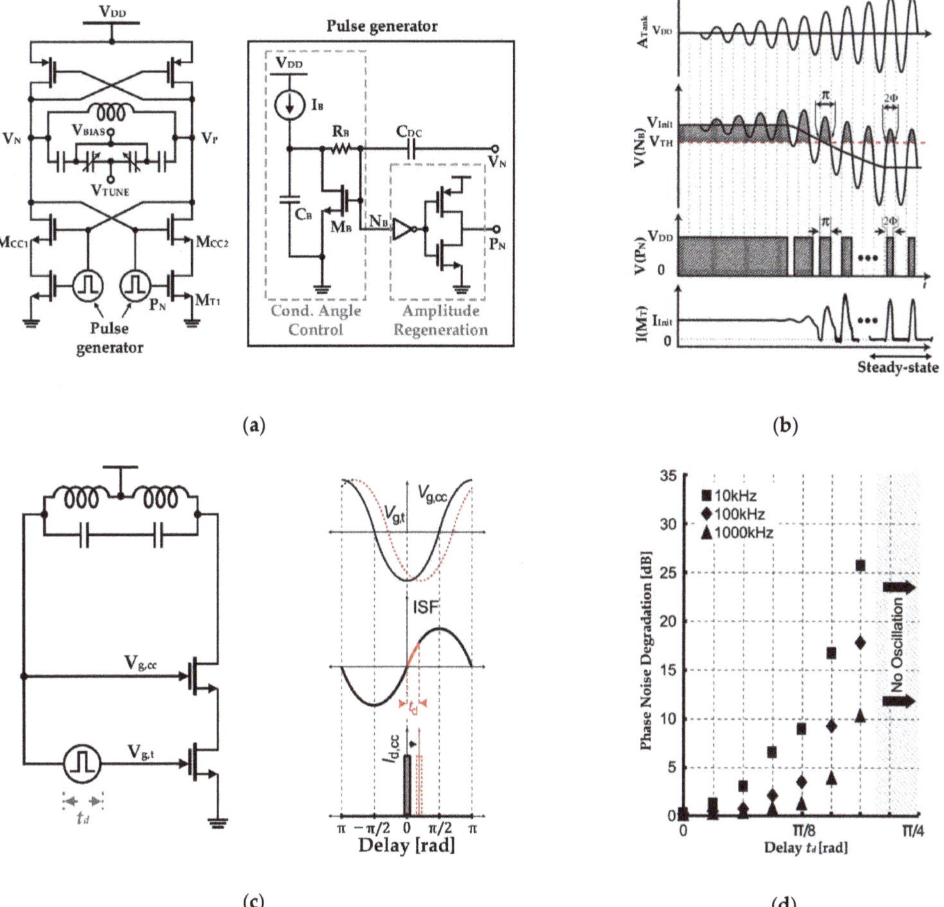

Figure 14. VCO employing the pulse generator that supplies pulse waveforms and the adaptive DC bias voltage: (**a**) schematic; (**b**) conceptual diagram of the VCO including the pulse generator operation; (**c**) intrinsic delay of the pulse generator; (**d**) simulated phase noise degradation with increasing the delay of the pulse generator (reproduced with permission from the author, A pulse-tail-feedback LC-VCO with 700 Hz flicker noise corner and −195dBc FoM; published by IEICE, 2019) [64].

4. Discussion

The switched-biasing technique in which the current source of the VCO is modulated with AC signals can reduce the closed-in phase noise of the VCO to minimize the generation of the flicker noise in the source. However, the effect on the overall performance of the VCO differs depending on the implementation method of generating the AC modulating signal. There are two general methods to generate the modulating signal: one uses the external signal generator and the other uses the oscillation signal coupled with the output of the VCO.

The external-biasing topology, which is a method using an external signal generator, showed that the closed-in phase noise of the VCO can be improved by the AC modulating signal of the current source [32,50]. However, it has drawbacks in the implementation of the integrated circuit because the chip area, power consumption, and design complexity can be increased by providing an external signal generator. In addition, as shown in Figure 15a, it is difficult to show the improvement in the performance caused by the reduction of the closed-in phase noise because the induced noise from an external generator can directly result in performance degradation of the VCO. The performance degradation caused by the induced noise from the external generator may be greater than the performance improvement caused by the reduction in the closed-in phase noise. The low correlation between the AC modulation signal and the oscillation signal may also increase the noise contribution of the external generator at the output of the VCO, as depicted in Figure 15b. It has been reported that the noise reduction of the VCO with an external generator is independent in the band below the modulation frequency, but the noise signals that are dependent on the modulation frequency are presented at the output of the VCO [42,49,50,56]. Based on the results of previous studies, it can be understood that the modulation frequency of the current source in the switched-biasing technique should be set to a frequency that does not affect the phase noise of the VCO. For example, the modulation frequency can be set to a frequency over the flicker corner frequency.

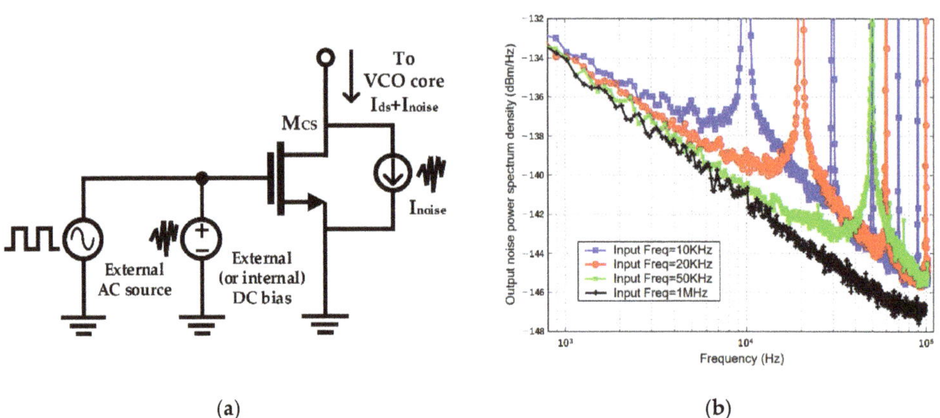

(a)   (b)

Figure 15. Characteristic of external-biasing topology: (**a**) conceptual schematic of external biasing (also available in PMOS configuration); (**b**) large noise peaks appearing as harmonics of the modulation frequency correlated with the external signal (reproduced with permission from the author, Experimental study on MOSFET's flicker noise under switching conditions and modelling in RF applications; published by IEEE, 2001) [56].

The self-biasing topology, which is a method using the coupled signal from the VCO output, has been proposed to solve the problem caused by the use of an external generator [34]. As the tail current source of the VCO in the self-biasing topology is modulated by the oscillation signal, the correlation between the modulation signal and the output can be achieved using easy implementation without additional circuitry. The current efficiency driving the VCO core can be increased by decreasing the conduction angle owing to the correlated AC modulation signal of the current source, based on the ISF theory [47,48]. In addition, the effect of the flicker noise of the current source is dramatically reduced in the self-biasing topology by the modulation frequency in the GHz band, which is higher than the flicker corner frequency. The DC bias voltage of the current source becomes an important design condition when the switched-biasing technique applies to the NMOS VCO using a low power supply voltage of 1.2 V or less. The efficiency of the switching operation is generally determined by the DC bias voltage, which is set near the threshold voltage because the current modulation should be exhibited by a small-sized switching signal. The DC bias of the current source in the initial research stage using the switched-biasing technique was set to the same biasing state as before the AC modulation of the source. However, the bias was changed to the threshold voltage of the current source transistor for clearly switching the on-off states of the current flow due to AC modulation [53,54,63]. The DC bias near the threshold voltage may not sufficiently supply the driving current for VCO operation, because the biasing current is generated at the sub-threshold region of the current source transistor, as shown in Figure 16a. In addition, it is difficult to apply to commercial circuits as operation reliability problems of the current source may occur owing to the PVT variation. Because the thermal noise produced from the voltage source for setting the DC bias can degrade the phase noise of the VCO, a method for reducing the contribution of the thermal noise should be applied to the circuit design for supplying the DC bias [58]. The capacitance of the resonator in the VCO is increased by the output coupling lines for self-biasing, as depicted in Figure 16b, and the resonance frequency of the LC tank can be affected by this increase. The coupling capacitor C_C in Figure 16b, which determines the amplitude of the modulation signal, should be generally designed to be higher than the parasitic capacitance of the current source [51,58]. The frequency shift due to the additional capacitances becomes an important factor in designing the high-frequency VCO because the total reactance in the LC tank decreases as the oscillation frequency increases. Above all, the major problem in DC biasing near the threshold voltage of the current source is that the current source does not operate in the saturation, and the common node of the VCO does not achieve a high impedance. Low impedance at the common node may cause the deduction of the phase noise as more noise from the switching operation of the current source affects the VCO core [58]. The effect on impedance reduction at the common node by DC biasing near the threshold voltage may be minimized as proposed by the previous studies, which include the method of splitting the current source into several transistors, the method of using a source degeneration capacitor, and the method of implementing an additional filter for noise reduction [52,54,58,60,64].

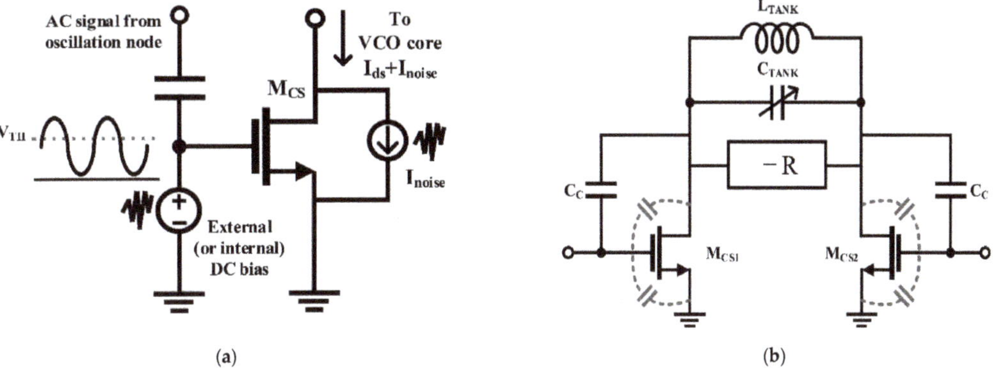

Figure 16. Characteristics of the self-biasing topology: (**a**) description of the self-biasing; (**b**) tuning range limit by the effect of the parasitic capacitances present at the current sources.

When the DC biasing of the current source is set to a voltage higher than the threshold voltage, the flicker noise may not be reduced by degrading the effect of the switching operation at the current source. An adaptive DC biasing technique that sets the biasing voltage differently depending on the amplitude of the oscillation signal has been proposed to reduce the operation problem generated by the fixed DC biasing at the current source. The adaptive biasing technique has the advantage of increasing the stability of the VCO operation and the robustness of the start-up operation. It was shown that the self-biasing topology implementing adaptive DC biasing using a negative peak detector can compensate for PVT variation and the variation of the Q-factor of the varactor that occurs in tuning the oscillation frequency [63]. A self-biasing topology with a pulse generator with three different operating states has been proposed to achieve a low conduction angle in steady state and a fast start-up time [64]. As the main drawback, the auxiliary circuit to implement the adaptive DC biasing requires additional loading to reduce the Q-factor of the LC tank and tuning range and increases the power consumption. The advantages and disadvantages of each topology implementing the switched-biasing technique are summarized in Table 1.

Table 1. Characteristics of different bias schemes.

Topologies	Bias Techniques	Advantages	Disadvantages
External-biasing	- The current source's gate is biased by external clock generator.	- Effectively reducing flicker noise generation.	- Requires circuit or equipment to generate ac signal. - Uncorrelated ac source noise directly degrades phase noise.
Self-biasing with the fixed DC voltage	- The current source's gate is coupled with the oscillation node of VCO. - DC bias is fixed to a specific value.	- No additional AC source required. - Circuit noise controlled internally	- Thermal noise is introduced through bias path. - Start-up issue
Self-biasing with the adaptive DC voltage	- The current source's gate is coupled with the oscillation node of VCO. - DC bias is controlled by the feedback path.	- VCO amplitude robust to PVT variation - Prevent oscillation start-up issue	- Complex circuitry - The adaptive DC biasing circuit is loaded directly to the VCO core, degrading performance.

The performances of the CMOS VCO using the switched-biasing technique are summarized in Table 2. The performances of VCOs oscillating at different frequencies are quantitatively compared using the conventional figure-of-merit (*FoM*) and the figure-of-merit with tuning range (*FoM$_T$*) as follows [65]:

$$FoM[dBc] = L(\Delta f) - 20\log\left(\frac{f_0}{\Delta f}\right) + 10\log\left(\frac{P_{DC}}{1mW}\right), \qquad (2)$$

$$FoMT[dBc] = L(\Delta f) - 20\log\left(\frac{f_0}{\Delta f}\right) + 10\log\left(\frac{P_{DC}}{1mW}\right) - 20\log\left(\frac{TR}{10}\right), \qquad (3)$$

where $L(\Delta f)$ is the phase noise of the VCO in dB at the frequency offset Δf from the center oscillation frequency, P_{DC} is the power consumption, and TR is the frequency tuning range in Hz. The phase noise, *FoM*, and *FoM$_T$* are normalized at a frequency offset of 1 MHz. In the ring-VCO, the self-biasing topology showed more improvement in the phase noise and *FoM* than the external-biasing topology. The LC-VCOs using the self-biasing topology showed a relatively high level of *FoM* below −179 dBc and FoM_T below −174 dBc. The LC-VCO with a *FoM* of −190 dBc showed a FoM_T of −179 dBc, which is a relatively low performance owing to the narrow tuning range [60]. The high *FoM* of −190 dBc is based on the phase noise reduction by the switched-biasing technique along with the reduction in the intrinsic noise and current consumption by using a PMOS current source operating in the triode mode [54,60]. The high *FoM* and FoM_T of the VCO with the adaptive DC voltage, high modulation amplitude, and pulse waveform in the self-biasing topology show that the minimum conduction angle can be useful for improving the VCO performance [64]. Figure 17 shows that the performance of the VCO can be improved by reducing the noise injection when the modulation waveform is implemented as a pulse with a duty cycle of less than π in the switched-biasing technique [64].

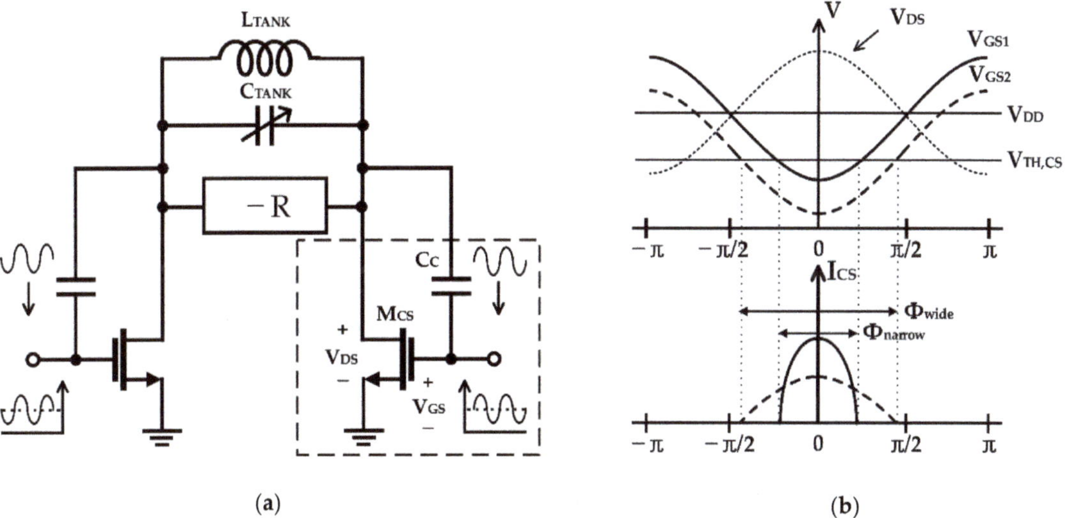

Figure 17. Conduction angle in the VCO operation: (**a**) description of the conduction angle in the VCO; (**b**) comparison of the conduction angle between the constant-biasing and the switched-biasing techniques.

Table 2. Performance summary of CMOS VCO using a switched-biasing technique.

Ref. (year)	Process (μm)	Bias Scheme	VCO Type	V_{DD} (V)	Freq. (GHz)	Tuning Range (%)	Phase Noise @1 MHz (dBc/Hz)	Power (mW)	FoM @1 MHz (dBc/Hz)	FoM_T @1 MHz (dBc/Hz)
[50] (2007)	0.18	External biasing	CMOS Ring VCO	1	1	82	−88 [1]	0.71	−149 [1]	−168 [1]
[61] (2020)	0.18	Self-biasing w. fixed DC	CMOS Ring VCO	1.8	1	47.6	−106	1.2	−165	−179
[51] (2011)	0.065	Self-biasing w. fixed DC	NMOS LC-VCO	1.2	20	17	−107	19.2	−179	−184
[58] (2015)	0.18	Self-biasing w. fixed DC	CMOS LC-VCO	1.2	2.55	9.2	−122.8	3.2	−186	−185
[60] (2015)	0.13	Self-biasing w. fixed DC	NMOS LC-VCO	1.4	2.4	1.7	−128.4	4.2	−190	−174
[54] (2019)	0.18	Self-biasing w. fixed DC	CMOS LC-VCO	1.2	2.45	28.6	−120	1.73	−185	−195
[53] (2019)	0.18	Self-biasing w. fixed DC	CMOS LC-VCO	0.8	1.4	18	−123	0.7	−187	−193
[57] (2006)	0.13	Self-biasing w. fixed DC	CMOS LC-QVCO	1.2	5	20	−117	5.28	−184	−190
[52] (2009)	0.18	Self-biasing w. fixed DC	CMOS LC-QVCO	1.8	2	17	−134.5	36	−185	−190
[63] (2009)	0.13	Self-biasing w. adaptive DC	NMOS LC-VCO	0.6	4.85	10.2	−117	3.9	−185	−185
[64] (2019)	0.18	Self-biasing w. adaptive DC	CMOS LC-VCO	1.2	4.55	4.3	−123.4	1.35	−195	−188

[1] Normalized at 1 MHz for comparison.

5. Conclusions

It was shown with physical and theoretical analyses that the switched-biasing technique can improve the phase noise characteristics by modulating the current source in the VCO using the deep sub-micron CMOS process. The switched-biasing technique can be divided into external-biasing and self-biasing topologies, depending on the method of implementing the current modulation. Even though the external-biasing topology can apply an optimum waveform as the modulation signal, the self-biasing topology that can control the current source with a waveform correlated with the output signal shows higher improvement in performance. The self-biasing topology can be subdivided into the usage of a fixed DC voltage and an adaptive DC voltage. The self-biasing topology with an adaptive DC voltage can be expected to apply the optimized waveform to the modulation signal; however, there is no significant improvement in the VCO performance compared to the self-biasing topology with a fixed DC voltage because the implementation of additional circuits, which are required for the adaptive DC voltage, increases the noise injection to the VCO. In addition, the self-biasing topology with the adaptive DC voltage is not also suitable for the millimeter-wave VCO design because the additional circuits can increase the parasitic components that affect the oscillation frequency shift, tuning range limit, and design accuracy. Based on the improvement of the phase noise, ease of implementation, and overall *FoM* and *FoM$_T$*, it could be concluded that the self-biasing topology with a fixed DC voltage is the most useful in the switched-biasing technique.

Author Contributions: Conceptualization, C.-W.K. and J.-R.Y.; methodology, C.-W.K.; software, C.-W.K.; validation, C.-W.K., H.M., and J.-R.Y.; formal analysis, C.-W.K. and J.-R.Y.; investigation, C.-W.K.; resources, C.-W.K. and H.M.; data curation, C.-W.K., H.M., and J.-R.Y.; writing—original draft preparation, C.-W.K., H.M., and J.-R.Y.; writing—review and editing, H.M. and J.-R.Y.; visualization, C.-W.K.; supervision, J.-R.Y.; project administration, J.-R.Y.; funding acquisition, J.-R.Y. All authors have read and agreed to the published version of the manuscript.

Funding: This work was partly supported by the Institute of Information & Communications Technology Planning & Evaluation (IITP) grant funded by the Korean government (MSIT) (No. 2018-0-00711) and the Basic Science Research Program through the National Research Foundation of Korea (NRF) funded by the Ministry of Education (No. 2017R1D1A1B03031357).

Institutional Review Board Statement: Not applicable.

Informed Consent Statement: Not applicable.

Data Availability Statement: No new data were created in this study. Data sharing is not applicable to this article.

Acknowledgments: The authors would like to thank all authors of previous papers for approving the use of their published research results in this paper.

Conflicts of Interest: The authors declare no conflict of interest.

References

1. Jann, B.; Chance, G.; Roy, A.G.; Balakrishnan, A.; Karandikar, N.; Brown, T.; Li, X.; Davis, B.; Ceballos, J.L.; Tanzi, N.; et al. 21.5 A 5G Sub-6GHz Zero-IF and mm-Wave IF Transceiver with MIMO and Carrier Aggregation. In Proceedings of the 2019 IEEE International Solid- State Circuits Conference—(ISSCC), San Francisco, CA, USA, 17–21 February 2019; pp. 352–354.
2. Khalaf, K.; Vaesen, K.; Brebels, S.; Mangraviti, G.; Libois, M.; Soens, C.; Thillo, W.V.; Wambacq, P. A 60-GHz 8-Way Phased-Array Front-End With T/R Switching and Calibration-Free Beamsteering in 28-nm CMOS. *IEEE J. Solid-State Circuits* **2018**, *53*, 2001–2011. [CrossRef]
3. Dunworth, J.D.; Homayoun, A.; Ku, B.; Ou, Y.; Chakraborty, K.; Liu, G.; Segoria, T.; Lerdworatawee, J.; Park, J.W.; Park, H.; et al. A 28GHz Bulk-CMOS Dual-Polarization Phased-Array Transceiver with 24 Channels for 5G User and Basestation Equipment. In Proceedings of the 2018 IEEE International Solid—State Circuits Conference—(ISSCC), San Francisco, CA, USA, 11–15 February 2018; pp. 70–72.
4. Angevare, J.; Pedalà, L.; Sönmez, U.; Sebastiano, F.; Makinwa, K.A.A. A 2800-μm^2 Thermal-Diffusivity Temperature Sensor with VCO-Based Readout in 160-nm CMOS. In Proceedings of the 2015 IEEE Asian Solid-State Circuits Conference (A-SSCC), Xiamen, China, 9–11 November 2015; pp. 1–4.

5. Enache, A.; Drăghici, F.; Pristavu, G.; Brezeanu, G. Voltage Controlled Oscillator for Small-Signal Capacitance Sensing. In Proceedings of the 2019 International Semiconductor Conference (CAS), Sinaia, Romania, 9–11 October 2019; pp. 323–326.

6. Quintero, A.; Cardes, F.; Perez, C.; Buffa, C.; Wiesbauer, A.; Hernandez, L. A VCO-Based CMOS Readout Circuit for Capacitive MEMS Microphones. *Sensors* **2019**, *19*, 4126. [CrossRef] [PubMed]

7. Vornicu, I.; Carmona-Galán, R.; Rodríguez-Vázquez, Á. Arrayable Voltage-Controlled Ring-Oscillator for Direct Time-of-Flight Image Sensors. *IEEE Trans. Circuits Syst. Regul. Pap.* **2017**, *64*, 2821–2834. [CrossRef]

8. Tseng, S.; Kao, Y.; Peng, C.; Liu, J.; Chu, S.; Hong, G.; Hsieh, C.; Hsu, K.; Liu, W.; Huang, Y.; et al. A 65-nm CMOS Low-Power Impulse Radar System for Human Respiratory Feature Extraction and Diagnosis on Respiratory Diseases. *IEEE Trans. Microw. Theory Tech.* **2016**, *64*, 1029–1041. [CrossRef]

9. Park, J.-H.; Yang, J.-R. Two-Tone Continuous-Wave Doppler Radar Based on Envelope Detection Method. *Microw. Opt. Technol. Lett.* **2020**, *62*, 3146–3150. [CrossRef]

10. Sim, J.Y.; Park, J.-H.; Yang, J.-R. Vital-Signs Detector Based on Frequency-Shift Keying Radar. *Sensors* **2020**, *20*, 5516. [CrossRef] [PubMed]

11. Bassi, M.; Caruso, M.; Bevilacqua, A.; Neviani, A. A 65-nm CMOS 1.75–15 GHz Stepped Frequency Radar Receiver for Early Diagnosis of Breast Cancer. *IEEE J. Solid-State Circuits* **2013**, *48*, 1741–1750. [CrossRef]

12. Tu, C.; Wang, Y.; Lin, T. A Low-Noise Area-Efficient Chopped VCO-Based CTDSM for Sensor Applications in 40-nm CMOS. *IEEE J. Solid-State Circuits* **2017**, *52*, 2523–2532. [CrossRef]

13. Peng, K.; Chen, S.; Wang, F.; Horng, T. Enhancement of Vital-Sign Sensor Signal-to-Noise Ratio Using Wireless Frequency-Locked Loop. In Proceedings of the 2019 IEEE SENSORS, Montreal, QC, Canada, 27–30 October 2019; pp. 1–3.

14. Pepe, F.; Bonfanti, A.; Levantino, S.; Samori, C.; Lacaita, A.L. Analysis and Minimization of Flicker Noise Up-Conversion in Voltage-Biased Oscillators. *IEEE Trans. Microw. Theory Tech.* **2013**, *61*, 2382–2394. [CrossRef]

15. Hu, Y.; Siriburanon, T.; Staszewski, R.B. A Low-Flicker-Noise 30-GHz Class-F$_{23}$ Oscillator in 28-nm CMOS Using Implicit Resonance and Explicit Common-Mode Return Path. *IEEE J. Solid-State Circuits* **2018**, *53*, 1977–1987. [CrossRef]

16. Tang, K.; Lou, L.; Guo, T.; Chen, B.; Wang, Y.; Fang, Z.; Yang, C.; Wang, W.; Zheng, Y. A 4TX/4RX Pulsed Chirping Phased-Array Radar Transceiver in 65-nm CMOS for X-Band Synthetic Aperture Radar Application. *IEEE J. Solid-State Circuits* **2020**, *55*, 2970–2983. [CrossRef]

17. Bennett, H.; Brederlow, R.; Costa, J.; Cottrell, P.; Huang, W.; Immorlica, A.; Mueller, J.-E.; Racanelli, M.; Shichijo, H.; Weitzel, C.; et al. Device and Technology Evolution for Si-Based RF Integrated Circuits. *IEEE Trans. Electron Devices* **2005**, *52*, 1235–1258. [CrossRef]

18. Razavi, B. *RF Microelectronics (2nd Edition) (Prentice Hall Communications Engineering and Emerging Technologies Series)*, 2nd ed.; Prentice Hall Press: Upper Saddle River, NJ, USA, 2011; ISBN 978-0-13-713473-1.

19. Nemirovsky, Y.; Corcos, D.; Brouk, I.; Nemirovsky, A.; Chaudhry, S. 1/f Noise in Advanced CMOS Transistors. *IEEE Instrum. Meas. Mag.* **2011**, *14*, 14–22. [CrossRef]

20. Leeson, D.B. A Simple Model of Feedback Oscillator Noise Spectrum. *Proc. IEEE* **1966**, *54*, 329–330. [CrossRef]

21. Bevilacqua, A.; Andreani, P. An Analysis of 1/f Noise to Phase Noise Conversion in CMOS Harmonic Oscillators. *IEEE Trans. Circuits Syst. Regul. Pap.* **2012**, *59*, 938–945. [CrossRef]

22. Hu, Y.; Siriburanon, T.; Staszewski, R.B. Intuitive Understanding of Flicker Noise Reduction via Narrowing of Conduction Angle in Voltage-Biased Oscillators. *IEEE Trans. Circuits Syst. II Express Briefs* **2019**, *66*, 1962–1966. [CrossRef]

23. Franceschin, A.; Andreani, P.; Padovan, F.; Bassi, M.; Bevilacqua, A. A 19.5-GHz 28-nm Class-C CMOS VCO, with a Reasonably Rigorous Result on 1/f Noise Upconversion Caused by Short-Channel Effects. *IEEE J. Solid-State Circuits* **2020**, *55*, 1842–1853. [CrossRef]

24. Mazzanti, A.; Andreani, P. Class-C Harmonic CMOS VCOs, With a General Result on Phase Noise. *IEEE J. Solid-State Circuits* **2008**, *43*, 2716–2729. [CrossRef]

25. Deng, W.; Okada, K.; Matsuzawa, A. Class-C VCO With Amplitude Feedback Loop for Robust Start-Up and Enhanced Oscillation Swing. *IEEE J. Solid-State Circuits* **2013**, *48*, 429–440. [CrossRef]

26. Hong, C.-H.; Wu, C.-Y.; Liao, Y.-T. Robustness Enhancement of a Class-C Quadrature Oscillator Using Capacitive Source Degeneration Coupling. *IEEE Trans. Circuits Syst. II Express Briefs* **2015**, *62*, 16–20. [CrossRef]

27. Hegazi, E.; Sjoland, H.; Abidi, A.A. A Filtering Technique to Lower LC Oscillator Phase Noise. *IEEE J. Solid-State Circuits* **2001**, *36*, 1921–1930. [CrossRef]

28. Murphy, D.; Darabi, H. 2.5 A Complementary VCO for IoE That Achieves a 195 dBc/Hz FOM and Flicker Noise Corner of 200 kHz. In Proceedings of the 2016 IEEE International Solid-State Circuits Conference (ISSCC), San Francisco, CA, USA, 31 January–4 February 2016; pp. 44–45.

29. Guo, H.; Chen, Y.; Mak, P.; Martins, R.P. 26.2 A 0.08 mm^2 25.5-to −29.9 GHz Multi-Resonant-RLCM-Tank VCO Using a Single-Turn Multi-Tap Inductor and CM-Only Capacitors Achieving 191.6 dBc/Hz FoM and 130 kHz 1/f3 PN Corner. In Proceedings of the 2019 IEEE International Solid- State Circuits Conference—(ISSCC), San Francisco, CA, USA, 17–21 February 2019; pp. 410–412.

30. Rael, J.J.; Abidi, A.A. Physical Processes of Phase Noise in Differential LC Oscillators. In Proceedings of the IEEE 2000 Custom Integrated Circuits Conference (Cat. No. 00CH37044), Orlando, FL, USA, 24 May 2000; pp. 569–572.

31. Shahmohammadi, M.; Babaie, M.; Staszewski, R.B. A 1/f Noise Upconversion Reduction Technique for Voltage-Biased RF CMOS Oscillators. *IEEE J. Solid-State Circuits* **2016**, *51*, 2610–2624. [CrossRef]

32. Gierkink, S.L.J.; Klumperink, E.A.M.; van der Wel, A.P.; Hoogzaad, G.; van Tuijl, E.A.J.M.; Nauta, B. Intrinsic 1/f Device Noise Reduction and Its Effect on Phase Noise in CMOS Ring Oscillators. *IEEE J. Solid-State Circuits* **1999**, *34*, 1022–1025. [CrossRef]
33. Klumperink, E.A.M.; Gierkink, S.L.J.; van der Wel, A.P.; Nauta, B. Reducing MOSFET 1/f Noise and Power Consumption by Switched Biasing. *IEEE J. Solid-State Circuits* **2000**, *35*, 994–1001. [CrossRef]
34. Boon, C.C.; Do, M.A.; Yeo, K.S.; Ma, J.G.; Zhang, X.L. RF CMOS Low-Phase-Noise LC Oscillator through Memory Reduction Tail Transistor. *IEEE Trans. Circuits Syst. II Express Briefs* **2004**, *51*, 85–90. [CrossRef]
35. Hooge, F.N. 1/f Noise. *Phys. BC* **1976**, *83*, 14–23. [CrossRef]
36. Keshner, M.S. 1/f Noise. *Proc. IEEE* **1982**, *70*, 212–218. [CrossRef]
37. Hooge, F.N. 1/f Noise Is No Surface Effect. *Phys. Lett. A* **1969**, *29*, 139–140. [CrossRef]
38. McWhorter, A.L. 1/f Noise and Related Surface Effects in Germanium. Ph.D. Thesis, Massachusetts Institute of Technology, Cambridge, MA, USA, 1955.
39. Kirton, M.J.; Uren, M.J. Noise in Solid-State Microstructures: A New Perspective on Individual Defects, Interface States and Low-Frequency (1/f) Noise. *Adv. Phys.* **1989**, *38*, 367–468. [CrossRef]
40. Bloom, I.; Nemirovsky, Y. 1/f Noise Reduction of Metal-oxide-semiconductor Transistors by Cycling from Inversion to Accumulation. *Appl. Phys. Lett.* **1991**, *58*, 1664–1666. [CrossRef]
41. Dierickx, B.; Simoen, E. The Decrease of "'Random Telegraph Signal'" Noise in Metal-oxide-semiconductor Field-effect Transistors When Cycled from Inversion to Accumulation. *J. Appl. Phys.* **1992**, *71*, 2028–2029. [CrossRef]
42. van der Wel, A.P.; Klumperink, E.A.M.; Gierkink, S.L.J.; Wassenaar, R.F.; Wallinga, H. MOSFET 1/f Noise Measurement under Switched Bias Conditions. *IEEE Electron Device Lett.* **2000**, *21*, 43–46. [CrossRef]
43. Kolhatkar, J.S.; Salm, C.; Knitel, M.J.; Wallinga, H. Constant and Switched Bias Low Frequency Noise in p-MOSFETs with Varying Gate Oxide Thickness. In Proceedings of the 32nd European Solid-State Device Research Conference, Firenze, Italy, 24–26 September 2002; pp. 83–86.
44. van der Wel, A.P.; Klumperink, E.A.M.; Kolhatkar, J.S.; Hoekstra, E.; Snoeij, M.F.; Salm, C.; Wallinga, H.; Nauta, B. Low-Frequency Noise Phenomena in Switched MOSFETs. *IEEE J. Solid-State Circuits* **2007**, *42*, 540–550. [CrossRef]
45. Koh, J.; Schmitt-Landsiedel, D.; Thewes, R.; Brederlow, R. A Complementary Switched MOSFET Architecture for the 1/f Noise Reduction in Linear Analog CMOS ICs. *IEEE J. Solid-State Circuits* **2007**, *42*, 1352–1361. [CrossRef]
46. Kim, J.; An, H.; Yun, T. A Low-Noise WLAN Mixer Using Switched Biasing Technique. *IEEE Microw. Wirel. Compon. Lett.* **2009**, *19*, 650–652. [CrossRef]
47. Soltanian, B.; Kinget, P.R. Tail Current-Shaping to Improve Phase Noise in LC Voltage-Controlled Oscillators. *IEEE J. Solid-State Circuits* **2006**, *41*, 1792–1802. [CrossRef]
48. Hajimiri, A.; Lee, T.H. A General Theory of Phase Noise in Electrical Oscillators. *IEEE J. Solid-State Circuits* **1998**, *33*, 179–194. [CrossRef]
49. van der Wel, A.P.; Klumperink, E.A.M.; Nauta, B. Measurement of MOSFET LF Noise under Large Signal RF Excitation. In Proceedings of the 32nd European Solid-State Device Research Conference, Firenze, Italy, 24–26 September 2002; pp. 91–94.
50. Yoshida, T.; Ishida, N.; Sasaki, M.; Iwata, A. Low-Voltage, Low-Phase-Noise Ring Voltage-Controlled Oscillator Using 1/f-Noise Reduction Techniques. *Jpn. J. Appl. Phys.* **2007**, *46*, 2257. [CrossRef]
51. Musa, A.; Murakami, R.; Sato, T.; Chaivipas, W.; Okada, K.; Matsuzawa, A. A Low Phase Noise Quadrature Injection Locked Frequency Synthesizer for MM-Wave Applications. *IEEE J. Solid-State Circuits* **2011**, *46*, 2635–2649. [CrossRef]
52. Huang, G.; Kim, B. Low Phase Noise Self-Switched Biasing CMOS LC Quadrature VCO. *IEEE Trans. Microw. Theory Tech.* **2009**, *57*, 344–351. [CrossRef]
53. Hsieh, J.; Lin, K. A 0.7-mW LC Voltage-Controlled Oscillator Leveraging Switched Biasing Technique for Low Phase Noise. *IEEE Trans. Circuits Syst. II Express Briefs* **2019**, *66*, 1307–1310. [CrossRef]
54. Shasidharan, P.; Ramiah, H.; Rajendran, J. A 2.2 to 2.9 GHz Complementary Class-C VCO with PMOS Tail-Current Source Feedback Achieving—120 dBc/Hz Phase Noise at 1 MHz Offset. *IEEE Access* **2019**, *7*, 91325–91336. [CrossRef]
55. Hsin-Shu, C.; Ito, A. Hsin-Shu Chen; Ito, A. Characterization of 1/f Noise vs. Number of Gate Stripes in MOS Transistors. In Proceedings of the 1999 IEEE International Symposium on Circuits and Systems (ISCAS), Orlando, FL, USA, 30 May–2 June 1999; Volume 2, pp. 310–313.
56. Zhang, Z.; Lau, J. Experimental Study on MOSFET's Flicker Noise under Switching Conditions and Modelling in RF Applications. In Proceedings of the IEEE 2001 Custom Integrated Circuits Conference (Cat. No.01CH37169), San Diego, CA, USA, 9 May 2001; pp. 393–396.
57. Jeong, C.; Yoo, C. 5-GHz Low-Phase Noise CMOS Quadrature VCO. *IEEE Microw. Wirel. Compon. Lett.* **2006**, *16*, 609–611. [CrossRef]
58. Chen, N.; Diao, S.; Huang, L.; Lin, F. Reduction of $1/f^3$ Phase Noise in LC Oscillator with Improved Self-Switched Biasing. *Analog Integr. Circuits Signal Process.* **2015**, *84*, 19–27. [CrossRef]
59. Tchamov, N.N.; Tchamov, N.T. Technique for Flicker Noise Up-Conversion Suppression in Differential LC Oscillators. *IEEE Trans. Circuits Syst. II Express Briefs* **2007**, *54*, 959–963. [CrossRef]
60. Mostajeran, A.; Bakhtiar, M.S.; Afshari, E. 25.8 A 2.4GHz VCO with FOM of 190 dBc/Hz at 10 kHz-to −2 MHz Offset Frequencies in 0.13 μm CMOS Using an ISF Manipulation Technique. In Proceedings of the 2015 IEEE International Solid-State Circuits Conference—(ISSCC) Digest of Technical Papers, San Francisco, CA, USA, 22–26 February 2015; pp. 1–3.

61. Lee, I.; Im, D. Low Phase Noise Ring VCO Employing Input-Coupled Dynamic Current Source. *Electron. Lett.* **2020**, *56*, 76–78. [CrossRef]
62. Hajimiri, A.; Limotyrakis, S.; Lee, T.H. Jitter and Phase Noise in Ring Oscillators. *IEEE J. Solid-State Circuits* **1999**, *34*, 790–804. [CrossRef]
63. Min, B.-H.; Hyun, S.-B.; Yu, H.-K. Low Voltage CMOS LC VCO with Switched Self-Biasing. *ETRI J.* **2009**, *31*, 755–764. [CrossRef]
64. Narayanan, A.T.; Okada, K. A Pulse-Tail-Feedback LC-VCO with 700Hz Flicker Noise Corner and -195dBc FoM. *IEICE Trans. Electron.* **2019**, *E102*, 595–606. [CrossRef]
65. Kinget, P. Integrated GHz Voltage Controlled Oscillators. In *Analog Circuit Design*; Sansen, W., Huijsing, J., van de Plassche, R., Eds.; Springer US: Boston, MA, USA, 1999; pp. 353–381. ISBN 978-1-4419-5101-4.

Article

An 18.8–33.9 GHz, 2.26 mW Current-Reuse Injection-Locked Frequency Divider for Radar Sensor Applications

Kwang-Il Oh [1], Goo-Han Ko [1], Jeong-Geun Kim [2] and Donghyun Baek [1,*]

[1] Microwave Embedded Circuit & System (MECAS) Lab., School of Electrical Engineering, Chung-Ang University, 84 Heukseok, Dongjack, Seoul 06974, Korea; dhrhkddlf123@cau.ac.kr (K.-I.O.); rngks79@cau.ac.kr (G.-H.K.)

[2] Integrated Radar Lab., Department of Electronic Engineering, KwangWoon University, 20 Gwangun, Nowon, Seoul 01897, Korea; junggun@kw.ac.kr

* Correspondence: dhbaek@cau.ac.kr; Tel.: +82-02-820-5828

Abstract: An 18.8–33.9 GHz, 2.26 mW current-reuse (CR) injection-locked frequency divider (ILFD) for radar sensor applications is presented in this paper. A fourth-order resonator is designed using a transformer with a distributed inductor for wideband operating of the ILFD. The CR core is employed to reduce the power consumption compared to conventional cross-coupled pair ILFDs. The targeted input center frequency is 24 GHz for radar application. The self-oscillated frequency of the proposed CR-ILFD is 14.08 GHz. The input frequency locking range is from 18.8 to 33.8 GHz (57%) at an injection power of 0 dBm without a capacitor bank or varactors. The proposed CR-ILFD consumes 2.26 mW of power from a 1 V supply voltage. The entire die size is 0.75 mm × 0.45 mm. This CR-ILFD is implemented in a 65 nm complementary metal-oxide semiconductor (CMOS) technology.

Keywords: current-reuse; injection-locked frequency divider; radar sensor; wideband

Citation: Oh, K.-I.; Ko, G.-H.; Kim, J.-G.; Baek, D. An 18.8–33.9 GHz, 2.26 mW Current-Reuse Injection-Locked Frequency Divider for Radar Sensor Applications. *Sensors* **2021**, *21*, 2551. https://doi.org/10.3390/s21072551

Academic Editor: Federico Alimenti

Received: 17 February 2021
Accepted: 2 April 2021
Published: 6 April 2021

Publisher's Note: MDPI stays neutral with regard to jurisdictional claims in published maps and institutional affiliations.

1. Introduction

Recently, the demand for radar sensors has been rapidly increasing with the development of the Internet of Things (IoT) industry and the autonomous vehicle industry. The complementary metal-oxide semiconductor (CMOS) radar is characterized by various operating methods such as doppler, frequency-modulated continuous wave (FMCW), and (continuous-wave) CW. In the doppler radar, a low frequency to millimeter-wave (mm-Wave) must be used to acquire a two-dimensional image through synthetic aperture radar (SAR). Bandwidths of 500 MHz or more are used to obtain high-resolution images [1,2]. In addition, wideband performance is very important in frequency-modulated continuous wave (FMCW) radars because wideband chirp is directly related to the high-resolution distance information [3]. Therefore, the wideband performance of the signal generator, the core of the sensor, is required [4,5].

Generally, the performance of the phase-locked loop (PLL) in the signal generator must be concerned to obtain low noise mm-Wave signals. Figure 1 shows the block diagram of conventional PLL structure that consists of a phase-frequency detector (PFD), charge pump (CP), low-pass filter (LPF), voltage-controlled oscillator (VCO) and frequency divider. The key blocks that determine the specification of the PLL in the mm-Wave band are VCO [6,7] and frequency divider [8]. The mm-Wave frequency divider should operate at high speed and should have a wide operating range for applying the wideband sensor applications.

Frequency dividers are designed as the current mode logic (CML) divider, regenerative divider, and LC oscillator-based injection-locked frequency divider (ILFD). The CML divider is a combination of two flip-flops that perform simple logical operations [9–12]. Generally, CML dividers have a wide operating range and occupy a small chip area with no inductor design. However, CML dividers suffer from large power consumption, limited maximum operation frequency, and process, voltage, and temperature (PVT) variation at the mm-Wave. To address these shortcomings, tunable self-resonant circuit [9], dynamic latches with load modulation [10,11], and additional calibration circuits [12] have been studied. However, these still consume large powers of 4.8 [11], and 6.2 mW [12], respectively. The regenerative divider and ILFD are also popular frequency dividers. These two types of frequency dividers are LC oscillator-based circuits and both of them are quite similar. The regenerative divider comprises an LC-based band pass filter (BPF) and active-type mixer [13–15]. The active type of mixer consumes power and takes over the role of -g_m core. Conversely, the ILFD comprises an LC-based BPF, -g_m core, and passive-type mixer that does not consume power. Therefore, regenerative dividers consume more power than ILFDs and are not generally used for mm-Wave applications because of the influence of many parasitic capacitors of the active-type mixer such as the Gilbert cell. The even-harmonic mixer [14] and digital-assisted circuit [15] are employed to widen the locking ranges of the regenerative divider. Their locking ranges are 33% and 57.4%, respectively. However, the highest input frequencies are limited to 18.4 and 14.8 GHz, consuming 10.8 and 12 mW power, respectively.

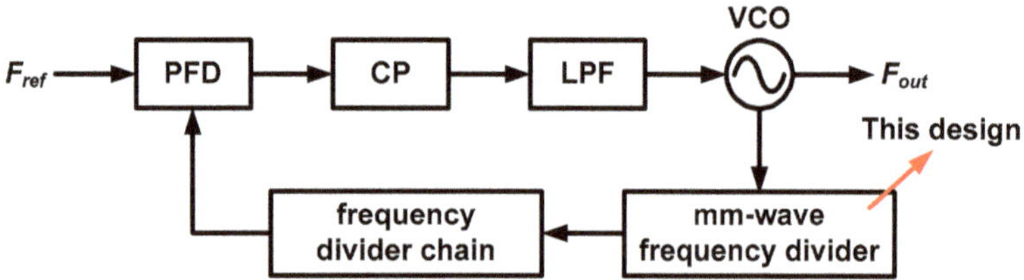

Figure 1. Conventional phase-locked loop with mm-Wave frequency divider.

The most attractive mm-Wave frequency divider is the LC oscillator-based ILFD. The reasons for its high popularity are as follows. First, the ILFD self-oscillates when there is no input signal applied. It is possible to obtain a large output signal with a small input signal using the oscillator-based operation. Second, because of the LC resonator, the ILFD is advantageous for operation at the mm-Wave band. Finally, because the ILFD uses a passive type of mixer, it consumes less power than regenerative and CML dividers. However, the disadvantage is that the locking range is narrow because of a high-quality factor (Q) LC resonator. Several studies are being conducted to widen the locking range of ILFD [16–18]. The forward-body-bias techniques [16,17] are some of the effective ways of increasing the gain of the mixer and extending the locking range. Although the ILFD with the forward-body-bias techniques have a wide locking range of 90% in [17], there are several reasons why this technique is impractical in mm-Wave synthesizers. First, if a positive bias is applied to the body of an n-channel metal-oxide-semiconductor field-effect transistor (MOSFET), the leakage current cannot be ignored, and the possibility of a large diffusion current flow because of forward-bias increases. Second, the power of the harmonic signal increases because of non-linearity in devices. Applying an injection signal with an edge frequency in the locked range can make it difficult to distinguish the power difference between the output and harmonic signals. Finally, an additional circuit may be required to control the harmonic power, which can increase the circuit complexity and power consumption. The dual-resonance resonator is also considered as a suitable technique [18]. This ILFD has a locking range of 71.46%; however, it requires external bias control and has a small output power. Moreover, when a −3 dBm injection power is applied, an unlocking part occurs in the locking range.

In this paper, a low power and wide locking range LC oscillator-based current-reuse (CR) ILFD using a fourth-order resonator with the distributed inductor is proposed. The CR technique is employed to reduce power consumption. This paper is organized as follows. Section 2 presents an analysis of the ILFD locking range. The limitations of the maximum locking range and harmonic issues are also presented. Section 3 presents the circuit design of the proposed CR-ILFD including the modeling of the transformer and design flow chart. The measurement results are shown in Section 4. Finally, conclusions are organized in Section 5.

2. Locking Range Analysis of ILFD

Figure 2a shows a schematic of the conventional cross-coupled pair ILFD with a second-order resonator. This ILFD consists of an N-channel metal-oxide semiconductor (NMOS) cross-coupled pair (M_1, M_2), injection switch (M_3) and LC resonator. The ILFD self-oscillates if there is no injection signal at the gate of M_3. Biasing the injection signal of $V_{inj,2w}$ at the gate of M_3, the ILFD outputs $V^+_{out,w}$ and $V^-_{out,w}$. When the frequency of the output signal is exactly half the frequency of the injection signal, it is referred to as "locking". To easily understand the locking operation, the current is classified into three types, namely, I_{so}, I_{inj}, and I_{out}. I_{so} represents the self-oscillation current flowing through the core when the ILFD self-oscillates without an injection signal. I_{inj} is the injection current flowing through M_3 when an injection signal is applied. I_{out} is the output current, which is the sum of I_{so} and I_{inj}. Figure 2b shows the phasor diagram for the three current types. The phasor rotates clockwise. Point "a" shows that the phase has changed from I_{so} by ϕ. Point "b" shows the phase when the ILFD self-oscillates without an injection signal. The relational expression of the current vectors is as follows.

$$I_{out} = I_{so} + I_{inj}. \tag{1}$$

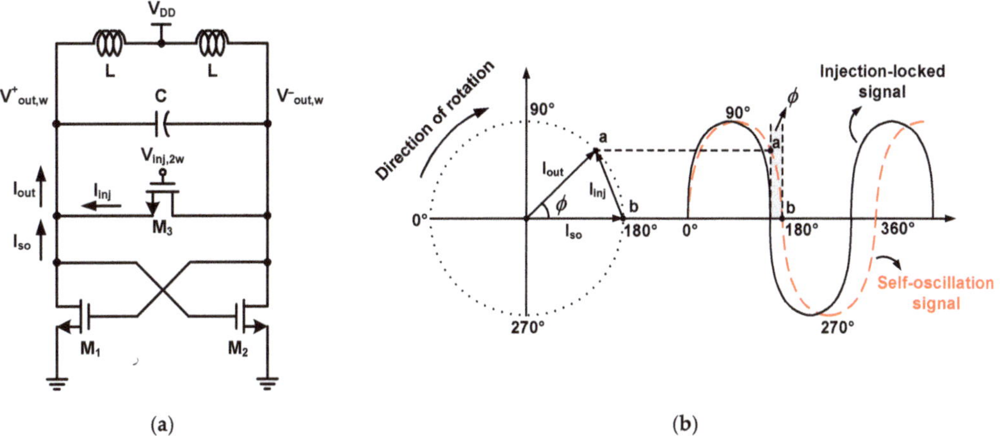

Figure 2. (**a**) Schematic of the conventional cross-coupled pair ILFD with second-order resonator and (**b**) phasor diagram for the basic principle of the conventional ILFD.

Two waves are shown in Figure 2b, one is the self-oscillation signal of the ILFD and the other is the injection-locked signal. Point "b" of the self-oscillation signal is moved to point "a" by the injection signal. Therefore, the phase at 180° of the injection-locked signal is point "a" of the self-oscillation signal. Injection is instantaneously performed every half period, and the range of ϕ can be derived using the following equations.

$$\phi = \angle I_{out} = \angle (I_{so} + I_{inj}), \tag{2}$$

$$V_{out} = Z_L \cdot I_{out}, \tag{3}$$

$$\angle I_{out} = \angle V_{out} - \angle Z_L, \tag{4}$$

where V_{out} is the output voltage signal when the ILFD is locked, and Z_L represents the load impedance of the LC resonator. Equation (4) can be derived using the phasor in (3).

To replace V_{out} with the self-oscillation and injection signals, the following equations are derived as

$$V_{out} = V_{so} + V_{inj},\qquad(5)$$

where V_{so} is the output voltage signal when the ILFD self-oscillates and V_{inj} is the injection voltage signal generated from M_3. It should be noted that V_{inj} is different from the input voltage signal, $V_{inj,2w}$. According to Equations (4) and (5), the ϕ is calculated as

$$\phi = \angle\left(V_{so} \pm V_{inj}\right) - \angle Z_L.\qquad(6)$$

The sign of V_{inj} is determined based on the value of the locked frequency relative to the self-oscillation frequency. When the ILFD self-oscillates with no injection signal, (6) is calculated as follows.

$$\phi|_{V_{inj}=0} = \angle V_{so} - \angle Z_L.\qquad(7)$$

V_{inj} is zero, and V_{so} is expressed as the product of I_{so} and Z_L. Because Z_L is canceled out, the following equation is satisfied:

$$\phi|_{V_{inj}=0} = \angle I_{so}.\qquad(8)$$

Meanwhile, ϕ_{max} is derived when the following condition is satisfied:

$$I_{out} \perp I_{inj}.\qquad(9)$$

The largest angle between I_{so} and I_{out} can be realized by considering the phasor as shown in Figure 2b. This is the condition of (9) where I_{out} and I_{inj} are vertical. Using the trigonometric function,

$$\sin\phi_{max} = \pm\frac{|I_{inj}|}{|I_{so}|},\qquad(10)$$

$$\phi_{max} = \pm\arcsin\left(\frac{|g_{inj} \cdot V_{inj}|}{|g_m \cdot V_{so}|}\right),\qquad(11)$$

where g_m and g_{inj} represent the transconductance of the cross-coupled pair and injection switch, respectively.

According to (6), the conditions for extending the locking range of the ILFD can be determined qualitatively. First, the magnitude of the self-oscillation signal V_{so} is decreased by reducing the sizes of M_1 and M_2 to decrease the transconductance of the cross-coupled pair. However, when the transconductance of the cross-coupled pair is too small, it can make failure in the self-oscillation, causing the ILFD to act as a harmonic buffer. Second, to increase the amplitude of V_{inj} generated by M_3, the size of M_3 can be increased or the injection signal $V_{inj,2w}$ can be amplified. However, the operation frequency may be limited by large parasitic capacitors. A pre-buffer, which consumes additional power, will be required to increase the amplitude of $V_{inj,2w}$. Finally, the phase of the load impedance can be changed. The phase of Z_L can increase or decrease ϕ. However, the maximum and minimum values of the phase, $\pm\phi_{max}$, limit the range of ϕ. Therefore, the phase of Z_L should be close to zero value in the wide frequency range. In conclusion, the maximum and minimum values of ϕ are determined by (11), and the method of extending the range of ϕ is consistent with the equation in (6).

The power of the output signal should be greater than that of the input signal. Two graphs of the load impedance magnitude against the angular frequency are shown in Figure 3, which presents two cases. The first case is the normal case where the power of the input signal is significantly smaller than that of the output signal as shown in Figure 3a. The range from w_1 to w_2 is the operation frequency band obtained by dividing by two, and the range from $2w_1$ to $2w_2$ is the injection frequency band. The operation and injection frequency bands do not overlap in the normal case because $2w_1$ is larger than w_2. Therefore, the input signal does not exceed the start-up condition and is not amplified more than the output signal. The second case is the abnormal case where the power of the input signal can be larger than that of the output signal, as in Figure 3b. Here, the operation and injection frequency bands overlap because $2w_1$ is smaller than w_2. The injection frequency band contains the parts that exceed the start-up conditions, which are determined by the following "Barkhausen formula".

$$g_m \cdot |Z_L| \geq 1. \tag{12}$$

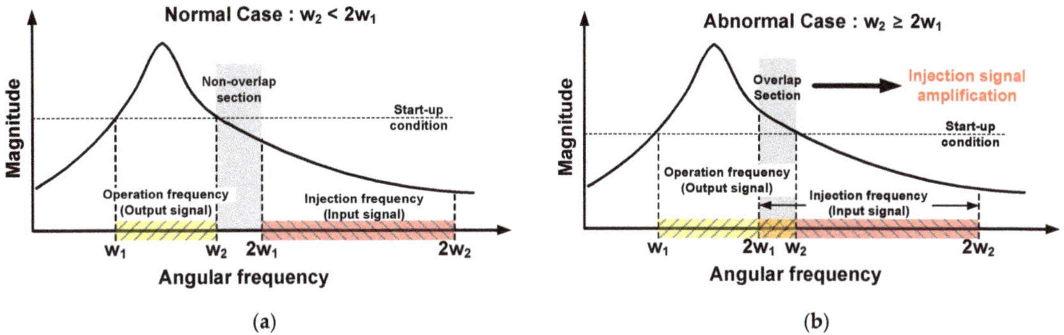

Figure 3. Graphs of magnitude of load impedance against angular frequency; (**a**) normal case, (**b**) abnormal case.

In the abnormal case, the ILFD cannot be used in mm-Wave applications, because the input and output signals are amplified together in the frequency band used.

This problem can be solved by increasing the division ratio of the ILFD. However, to operate at high division ratio, a harmonic signal with a small magnitude should be used, which results in a narrow locking range of the ILFD [19,20]. Additionally, the injection mixer for the high division ratio creates larger parasitic capacitance than the injection switch of the divide-by-two ILFD. Consequently, an ILFD that operates at a high division ratio greater than two is disadvantageous for application in the mm-Wave band. Therefore, a divide-by-two ILFD optimized to have a wide locking range without including the abnormal case would be most suited as a mm-Wave frequency divider. The following equation is used to calculate the locking range of the ILFD.

$$LR = \frac{w_2 - w_1}{w_1 + (w_2 - w_1)/2} \cdot 100 \ (\%). \tag{13}$$

Under the normal case condition, $w_2 < 2w_1$, the maximum locking range of the divide-by-two ILFD can be obtained when w_2 is equal to $2w_1$. Therefore, the maximum locking range is

$$LR_{\max}|_{w_2=2w_1} = 66.7\%, \tag{14}$$

where LR is the locking range. If the locking range of the divide-by-two ILFD exceeds 66.7%, the power of the input signal may be greater than that of the output signal. In conclusion, the locking range of the ILFD should be designed to be less than 66.7%.

3. Circuit Design of Proposed CR-ILFD

3.1. Fourth-Order Resonator and CR Core

As mentioned in the previous section, to extend the locking range of the ILFD, the phase plot of the load impedance should be flat in the range of $\pm\phi_{\max}$ [8,21]. A fourth-order resonator with two poles is required to flatten the phase plot. Figure 4a shows a schematic of the conventional cross-coupled pair-based ILFD with a fourth-order resonator consisting of a resonator ($L_1, C_1, R_1, L_2, C_2, R_2$), cross-coupled pair ($M_1, M_2$) and injection switch ($M_3$).

The "k" is the coupling factor between L_1 and L_2. Z_L is the load impedance of the resonator, which is calculated as

$$Z_L = \frac{(1-k^2)L_1L_2C_2s^3 + L_1s}{(1-k^2)L_1L_2C_1C_2s^4 + (L_1C_1 + L_2C_2)s^2 + 1}. \tag{15}$$

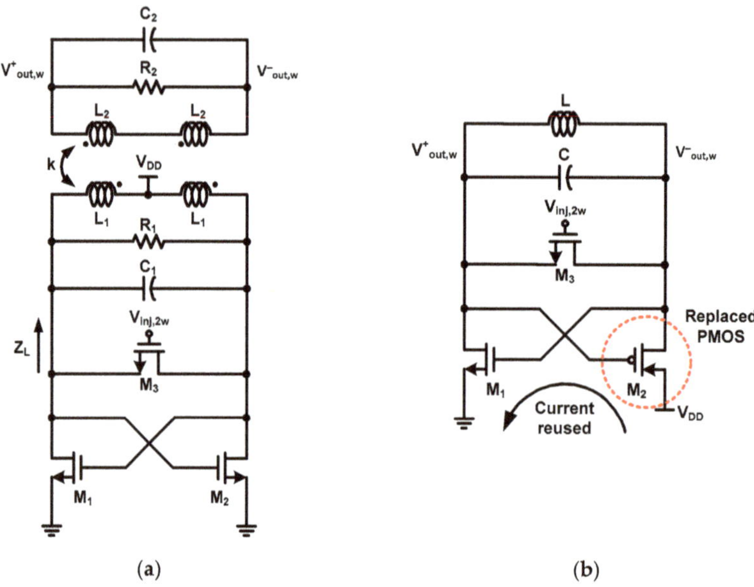

Figure 4. Schematic of (**a**) conventional cross-coupled pair ILFD with fourth-order resonator and (**b**) CR core-based ILFD.

R_1 and R_2 are resistors that affect the quality (Q) factor of the resonator and have been approximated in this calculation. Two poles that make the denominator zero are represented using the following equation [22],

$$w_{R,L} = \sqrt{\frac{L_1C_1 + L_2C_2 \pm \sqrt{(L_1C_1 + L_2C_2)^2 - 4(1-k^2)L_1L_2C_1C_2}}{2(1-k^2)L_1L_2C_1C_2}}. \tag{16}$$

Assuming that $L_1 = L_2$ and $C_1 = C_2$,

$$w_{L,R} = \frac{1}{\sqrt{(1 \pm k)LC}}. \tag{17}$$

According to (17), the distance between the two poles increases as the value of k increases and the distance between the two poles decreases as the k value decreases. If k is zero, the pole value is obviously equal to that of the second-order resonator (18). Figure 4b shows a schematic of the conventional ILFD with the CR core. For the CR core, M_2 of the cross-coupled pair ILFD in Figure 2a is replaced by P-channel metal-oxide semiconductor (PMOS) [23–27]. The oscillation of the CR core can be divided into two half periods. In the first half period, the current flows through M_1 and M_2, and in the second half period, no current flows through M_1 and M_2. Unlike the oscillation in the cross-coupled pair core, the oscillation of the CR core reduces the current by simultaneously turning the MOSFET on and off [24].

Figure 5 shows the magnitude and phase plots of the second- and fourth-order resonator-based ILFDs. The schematic of the second-order resonator-based ILFD is shown in Figure 2a. Figure 5a shows the graph of the load impedance magnitude against the input frequency. The second-order resonator-based ILFD has one pole, w_0, that is expressed as follows.

$$w_0 = \frac{1}{\sqrt{LC}}. \tag{18}$$

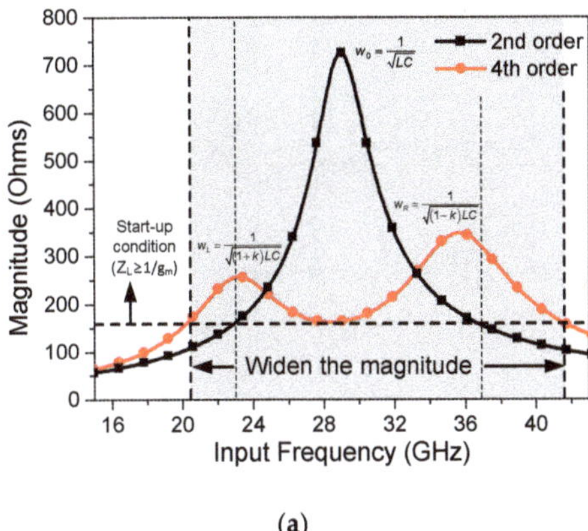

(a)

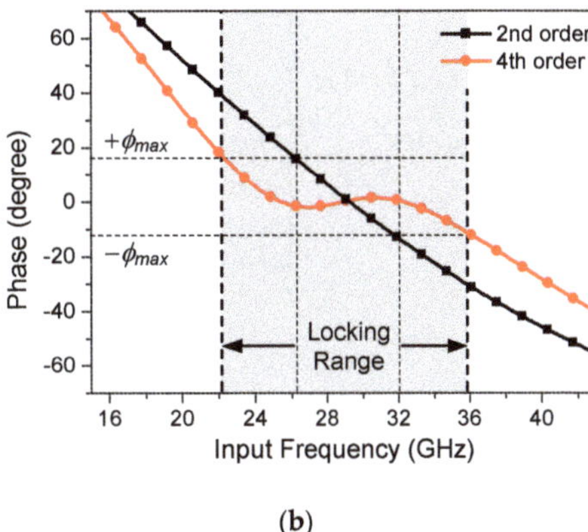

(b)

Figure 5. (**a**) Simulated magnitude plot and (**b**) phase plot of second-order resonator-based ILFD and fourth-order resonator-based ILFD.

If the fourth-order rather than the second-order resonator-based ILFD is applied, the magnitude plot of the load impedance becomes wider even if the maximum magnitude value decreases. However, because a new minimum value occurs between the two poles, it is necessary to simulate whether locking is sufficiently achieved at this value. If the minimum value between the two poles is less than the start-up condition (12), the ILFD does not operate in that frequency range. Figure 5b shows the phase plot against the input frequency. According to (11), the $\pm\phi_{max}$ limits the locking range of the ILFD. Unlike the phase of the second-order resonator-based ILFD, that of the fourth-order resonator-based ILFD has a value approximately equal to zero over a wide frequency range because of the formation of a ripple. Consequently, the simulated locking range of the ILFD is increased by 22% from 26–32 GHz (21%) to 22–36 GHz (43%).

3.2. Proposed CR-ILFD

Figure 6 shows a schematic of the proposed CR-ILFD, consisting of a fourth-order resonator (L_1, C_1, L_2, C_2), distributed inductor (L_3), injection switch (M_3), CR core (M_1, M_4), center-tap generator (M_2, M_5), and output buffer. $V_{inj,DC}$ and $V_{inj,2w}$ are the input signals, whereas $V_{out,w}$ is the output signal. The center-tap generator biases the node of the primary coil, L_1 to V_{CT}. If the g_m matching of PMOS and NMOS is well adjusted, mathematically, V_{CT} would be $V_{DD}/2$. The DC value of the injection switch can be biased to V_{CT} without additional supply, but it was not connected for measurement. The distributed inductor is employed to extend the locking range of the ILFD. The distributed inductor is also referred to as the inductor distributed technique [28,29]. The magnitude of the load impedance can be increased by distributing the primary inductor into two series inductors.

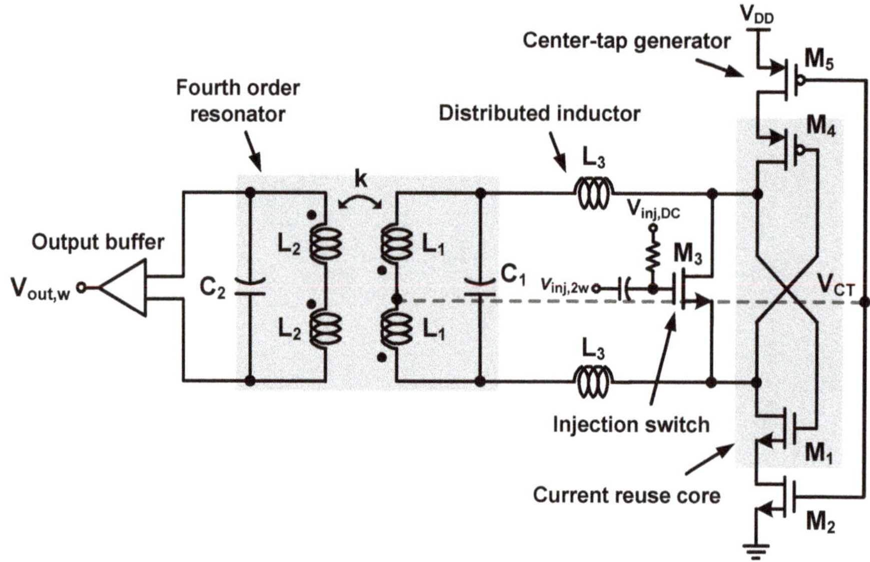

Figure 6. Schematic of the proposed CR-ILFD.

Figure 7a shows the simulated magnitude plot and phase plot of the fourth-order resonator-based ILFD and the proposed CR-ILFD with the fourth-order resonator with a distributed inductor. The start-up condition in Figure 7a is determined by the "Barkhausen formula" in (12). In the case of the fourth-order resonator-based ILFD, an unlocking part may occur because of the minimum value that is less than the start-up condition. However, the magnitude of the load impedance is sufficiently increased by using the inductor distributed technique. Figure 7b shows the slightly increased phase. This is not a critical amount of change because the phase ripple still exists between the $\pm\phi_{max}$. The simulated locking range of the proposed ILFD is from 21.6 to 37.4 GHz, which is limited by the $\pm\phi_{max}$ in (11).

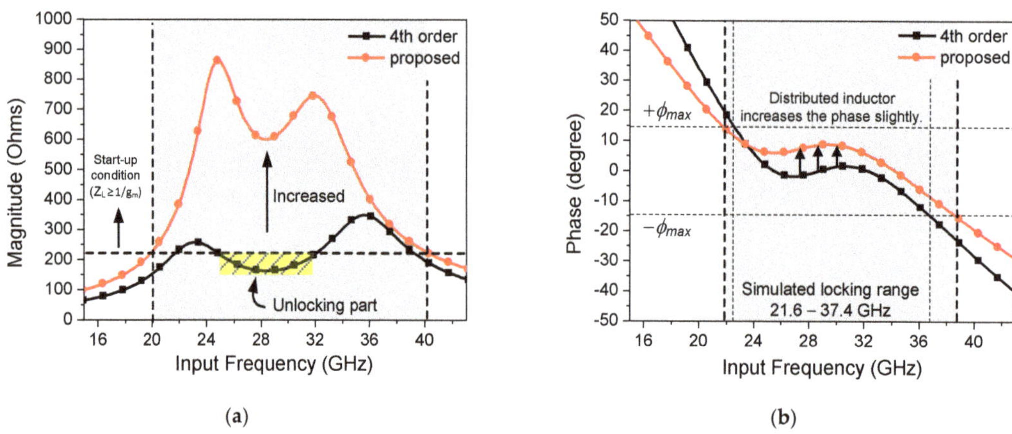

Figure 7. (**a**) Simulated magnitude plot and (**b**) phase plot of the fourth-order resonator-based ILFD and proposed CR-ILFD.

Figure 8 shows an equivalent model of the fourth-order resonator using a transformer with the distributed inductor. Figure 8a shows a model including the parasitic capacitors and resistors of the passive components. Zin is the input impedance and "*k*" is the coupling factor between L_1 and L_2. C_{p1}, R_{p1}, C_{p2}, and R_{p2} represent the parasitic components. In the mm-Wave band, the analog circuits are affected more by electromagnetism. Therefore, the modeling of the resonator must be considered at the initial design stage. Because analyzing every parasitic component is difficult, modeling should be simplified by approximation as shown in Figure 8b. C_{T1} is the sum of C_{p1} and C_1. Similarly, C_{T2} is the sum of C_{p2} and C_2. Additionally, the Q factor of the inductor includes the parasitic resistances. V_t and I_t are the test voltage and test current, respectively. V_t/I_t is equal to Z_{in} in the simplified model. The value of Z_{in} is calculated as follows.

$$Z_{in}(s) = \frac{(1-k^2)L_1L_2C_{T2}s^3 + L_1s}{(1-k^2)L_1L_2C_{T1}C_{T2}s^4 + (L_1C_{T1} + L_2C_{T2})s^2 + 1} \times (1 + 2L_3C_{T1}s^2). \qquad (19)$$

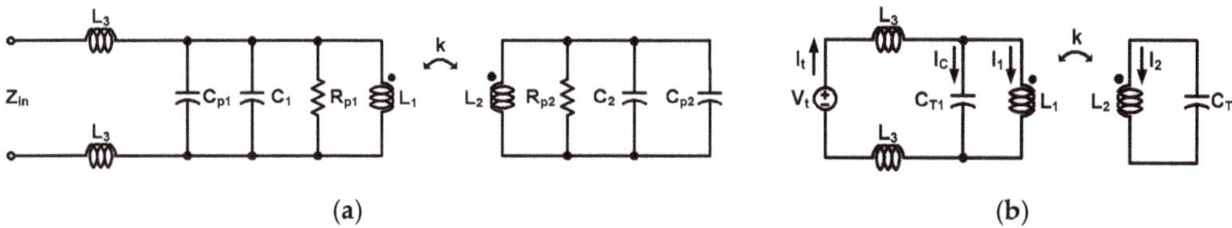

(a)   (b)

Figure 8. (a) Modeling of the fourth-order resonator using a transformer with distributed inductor. (a) Modeling of including the parasitic capacitors and resistors. (b) Approximate modeling applied to simplify calculations.

If the distributed inductor (L_3) is zero, then (19) is equal to (15). That is, the distributed inductor does not directly affect the pole value in (17), and if the distributed inductor value is increased, the magnitude of Z_{in} can be increased.

The design parameters are listed in Table 1. Because the center-tap generator should not limit the core operation, the width of the center-tap generator should be significantly larger than that of the CR core. The parasitic capacitor of the center-tap generator is separated from the resonator and does not affect the operating frequency. The sizes of the CR core and injection switch are not only determined by (11) and (12), but also by the influence of the parasitic capacitors.

Table 1. Design parameters of the proposed CR-ILFD.

Design Parameter	Value
M_1, M_2, M_4, M_5 (unit W/L)	2 μm/0.06 μm
M_3 (unit W/L)	1 μm/0.06 μm
Finger of M_1, M_3, M_4	20
Finger of M_2, M_5	50
L_1	230 pH
L_2	265 pH
L_3	433 pH
k	0.51
C_1	144 fF
C_2	240 fF

Figure 9 shows a flowchart of the design approach for the proposed CR-ILFD. First, the equivalent circuit model must be implemented in the simulator. Second, the values of the design parameters should be determined. In the proposed CR-ILFD, the center frequency is set to receive an injection signal of 28 GHz. Because the distributed inductor does not directly affect the pole value, the values of L_1, C_1, L_2, C_2 and k are first determined. Subsequently, L_1 is divided into two series inductors, L_1 and L_3. In this design, L_1, L_2, and L_3 are 230, 265, and 433 pH, respectively. C_1 and C_2 are 144 and 240 fF, respectively. The value of k is 0.51. When $k < 0.5$, which represents a weak coupling, the distance between the poles increases, and a wide magnitude plot of the load impedance can be obtained. However, a coupling that is too weak can cause an unlocking part in which the ILFD does not work. Considering the locking range and unlocking part, the proposed CR-ILFD is designed with a coupling factor of 0.51. Finally, the layout and locking simulation are repeated in the order shown in the flowchart. Electromagnetic simulation is essential in the mm-Wave band. Therefore, it should be ensured that the difference between the equivalent modeling and implementation in the simulation of this circuit is reasonable.

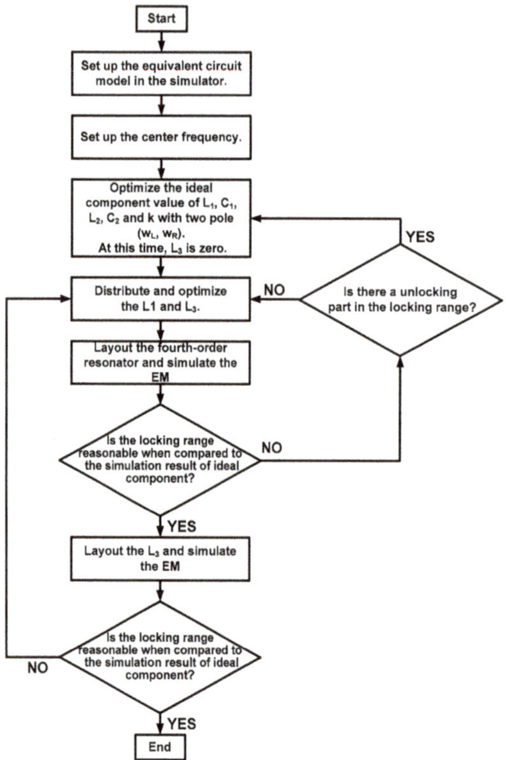

Figure 9. Flowchart of the design approach for the proposed CR-ILFD.

4. Measurement Results

Figure 10 shows the die photograph of the proposed CR-ILFD, which was fabricated in a 65 nm CMOS technology. The die size including the entire pad is 0.75 mm × 0.45 mm and the chip size including the core and output buffer is 0.49 mm × 0.3 mm. The measurement setup for the proposed CR-ILFD is shown in Figure 11. The measurements were obtained using a probe station. The DC voltage was biased from the power supply. The CR core of the proposed CR-ILFD consumes 2.26 mW from a 1 V supply voltage, when no signal is applied to the injection switch. As $V_{inj,DC}$ increases, the power consumption increases. When $V_{inj,DC}$ is 0.7 V, the power consumption of the core increases by approximately 0.5 mW. The power consumption of the output buffer is approximately 3 mW. The injection signal was generated by Anritsu MG3694, which can generate frequencies up to 40 GHz. The output signal of the proposed CR-ILFD is analyzed by KEYSIGHT N9030B, which can analyze frequencies up to 50 GHz. When conducting measurements using mm-Wave signals, several losses occur around the device under test (DUT). Therefore, the calibration tests must be carried out carefully. In this measurement, the ground–signal–ground (GSG) probe tip has a loss of approximately 2.5 dB and that of the radio frequency (RF) cable has approximately 3 dB. Approximately a 1 dB loss occurs even when the signal generator output is 10 dBm. The loss of the signal generator was analyzed by connecting the signal analyzer and RF cable. All losses described above are based on the 28 GHz signal. Generally, the loss increases as the frequency increases, and decreases as the frequency decreases.

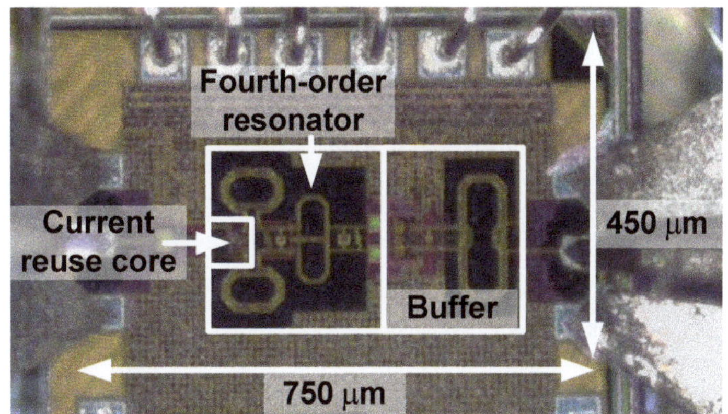

Figure 10. Die photograph of the proposed CR-ILFD.

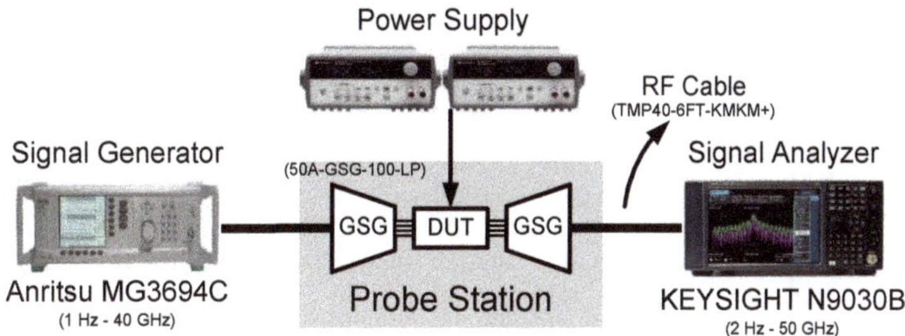

Figure 11. Measurement setup for the proposed CR-ILFD.

Figure 12a shows the measured locking range of the proposed CR-ILFD with different $V_{inj,DC}$ values. The maximum locking range is from 18.8 to 33.8 GHz (57%) at $V_{inj,DC}$ of 0.7 V. When the $V_{inj,DC}$ is biased to 0.6 V, the locking range is from 19.2 to 34.4 GHz (56.7%), and when the $V_{inj,DC}$ is biased to 0.5 V, the locking range is reduced from 22.7 to 34.6 GHz (41.5%). The above ranges were obtained from 0 dBm input power and 1 V supply voltage. As the $V_{inj,DC}$ decreases, the locking range also tends to decrease. Figure 12b shows a comparison of the measured and simulated locking range results of the proposed CR-ILFD. The measured locking range is 57%, and simulated locking range is from 21.6 to 37.4 GHz (53.6%). When 0 dBm input power is injected to the CR-ILFD, the measured locking range is typically changed to a lower frequency band than the simulated locking range. The operating frequency band was lowered by approximately 3 GHz. This is because of various electromagnetic components, such as RF pads, printed circuit board (PCB), and metal lines that were not considered in the simulations. The measured maximum operation frequency is higher when the input power is −3 dBm compared to when the input power is 0 dBm. This is because of the saturation of the input signal level.

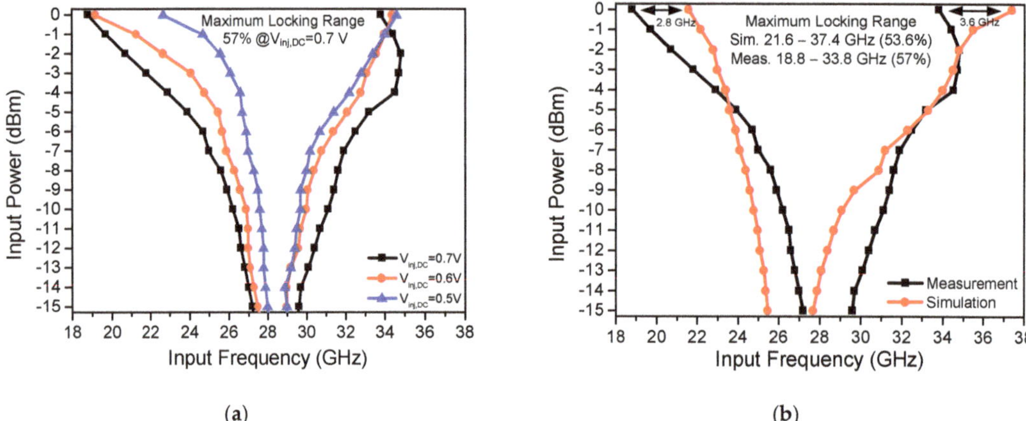

Figure 12. (**a**) Measured locking range results of the proposed CR-ILFD with different $V_{inj,DC}$; (**b**) measured and simulated locking range results of the proposed CR-ILFD.

Figure 13a shows the measured maximum and minimum operation frequencies of the proposed CR-ILFD with different $V_{inj,DC}$ values. This measurement was carried out with 0 dBm input power and 1 V supply voltage. $V_{inj,DC}$ is swept from 0.4 to 1.2 V, and the widest locking range is obtained at the $V_{inj,DC}$ of 0.7 V. When $V_{inj,DC}$ increases from 0.7 V, the maximum and minimum operation frequencies decrease, and the locking range also decreases.

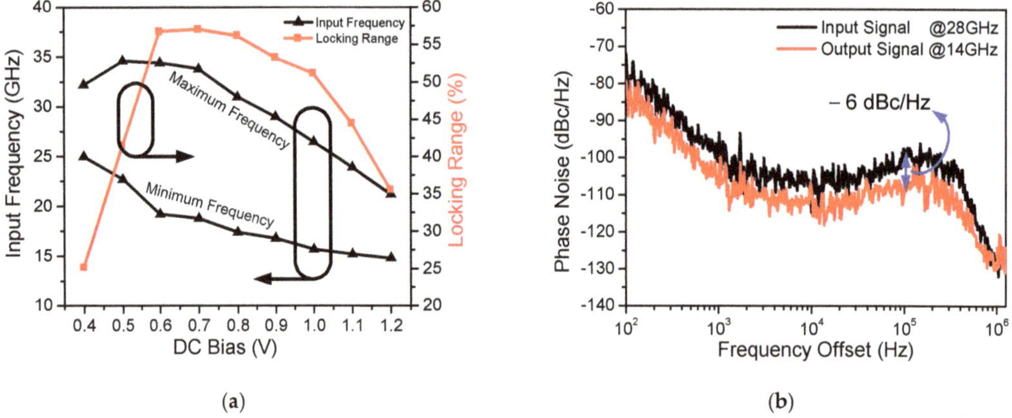

Figure 13. (**a**) Measured maximum and minimum operation frequency of the proposed CR-ILFD with different $V_{inj,DC}$; (**b**) Measured phase noise of input and output signal.

The measured phase noise of the input and output signal is shown in Figure 13b. The 28 GHz input signal is generated by Anritsu MG3694, which is applied to the proposed CR-ILFD and the output signal is 14 GHz. The phase noise of the output signal is -109.57 and -129.81 dBc/Hz at 100 kHz and 1 MHz offset frequency, respectively. The phase noise of the output signal should be measured at 6 dBc/Hz lower than that of the input signal because the input signal frequency is twice that of the output signal. Figures 14 and 15 show the results of several spectrums of the CR-ILFD's output signal measured using the KEYSIGHT N9030B. The spectrum of the output signal when the proposed CR-ILFD self-oscillates is shown in Figure 14a. The output frequency is 14.08 GHz, and output power is −10.45 dBm. If the loss of the RF cable and GSG probe tip is calibrated, the output power will be approximately −5 dBm. The spectrum of the output signal when the 28 GHz input signal is injected to the proposed CR-ILFD is shown in Figure 14b. The frequency of the output signal is 14 GHz, which is exactly half the frequency of the input signal. The output power is approximately −8 dBm with loss calibration. Figure 15a shows the full span spectrum when the minimum input frequency, 18.8 GHz, is injected. Three tones are visible in the spectrum: the output signal (f_0), input signal ($2f_0$), and harmonic signal ($3f_0$). As shown in Figure 3, several harmonic components are amplified at output when the minimum input frequency is injected to the CR-ILFD. Locking is possible even if a lower input frequency is injected. However, the input signal is amplified such that the power difference from the output signal is less than 10 dB. When 18.8 GHz is injected, the power difference between the desired output signal and the harmonic signal is approximately 10 dB. Figure 15b shows the full span spectrum when the maximum frequency input signal of 33.8 GHz is injected. The power difference between the output and input signals is more about 20 dB. It can be observed that the amplified input signal is smaller when the maximum input frequency is injected than when the minimum input frequency is injected. As a result, harmonic rejection ratio of the input signal over the entire locking range is more than 10 dBc.

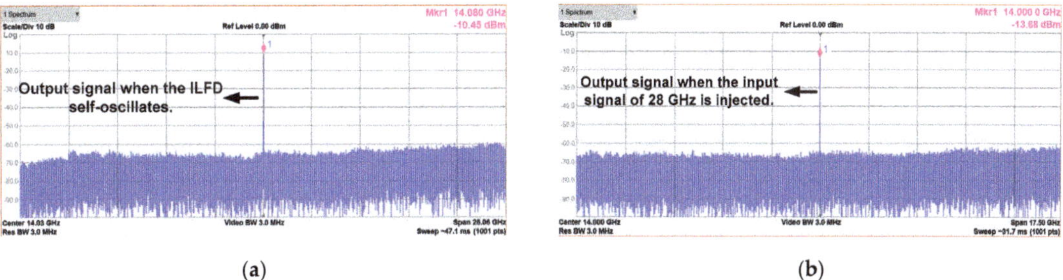

(a)   (b)

Figure 14. Spectrums of the output signal (**a**) when the proposed CR-ILFD self-oscillates; (**b**) when the proposed CR-ILFD is locked with a 28 GHz injection signal.

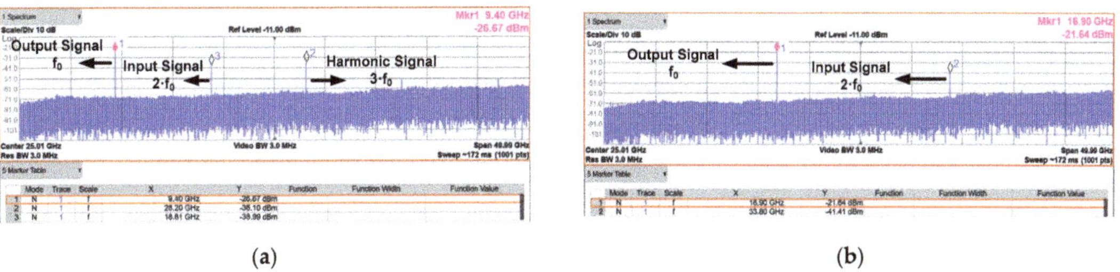

(a)   (b)

Figure 15. Full span spectrums (**a**) when the minimum input frequency is injected (18.8 GHz); (**b**) when the maximum input frequency is injected (33.8 GHz). The power difference between the output and input signals is approximately 10 dB or more.

Table 2 summarizes the performance comparison of different core ILFDs. These include challenging and typical ILFD cores such as Darlington [30], Armstrong [31], Collpits [32], and cross-coupled pair [33–35]. This work has the highest figure of merit (FOM) compared to other ILFDs presented in Table 2.

Table 2. Performance comparison of different core ILFDs.

	This Work	[30] 15′MTT	[31] 09′MWCL	[32] 08′MWCL	[33] 14′MWCL	[34] 15′APMC	[35] 17′MWCL
Technology	65-nm CMOS	0.18-μm CMOS	0.18-μm CMOS	0.18-μm CMOS	0.18-μm SiGe BiCMOS	0.18-μm CMOS	0.18-μm CMOS
Core topology	Current reuse	Darlington	Armstrong	Colpitts + Current reuse	NMOS Complementary cross-coupled pair	NMOS cross-coupled pair	NMOS cross-coupled pair
Self-oscillation frequency (GHz)	14.08	N/A	4.77–5.08 (w/varactor)	5.85–6.17 (w/varactor)	N/A	N/A	2.97–4.66 (w/varactor)
Input signal power (dBm)	0	0	0	0	0	0	0
Division ratio	2	2	2	2	2	2	4
Input frequency range (GHz)	18.8–33.8	20.5–22.9	7.7–11.5	7.3–14.4	20.1–25.9	10.2–15.5	13–19
Locking range (%)	57	11	39.6	65.4	25.1	41.4	37.5 *
Supply voltage (V)	1	1.2	1.4	1.5	1.8	1.2	0.8
Power consumption of core (mW)	2.26	1.73	9.02	7.65	4.8	2.71	7.09
Phase noise (dBc/Hz @1 MHz)	−129.81 (14 GHz)	−138.3 (N/A)	−134.942 (4.9 GHz)	−134.8 (6 GHz)	−124 (12.5 GHz)	−120.53 (5.495 GHz) @ 100 kHz	−133.26 (4 GHz)
FOM_1 (GHz/mW)	6.64	1.38	0.42	0.93	1.21	1.96	0.85
FOM_2 (GHz/mW)	13.28	2.76	0.84	1.86	2.42	3.92	3.4
Chip size (mm^2)	0.75 × 0.45	0.8 × 0.75	0.55 × 0.74	0.46 × 0.52	0.75 × 0.78	0.57 × 0.68	1.01 × 1.18

FOM_1 = Input frequency range/power consumption [GHz/mW], FOM_2 = (Input frequency range × division ratio)/power consumption [GHz/mW], *: Total locking range (low band + high band).

Table 3 summarizes the performance comparison of the mm-Wave ILFDs [36–41]. ILFDs with division ratio greater than two are also included such as four [37,38] and six [41], but still have the highest FOM$_1$ values.

Table 3. Performance comparison of mm-Wave ILFDs.

	This Work	[36] 15′MWCL	[37] 13′TCAS1	[38] 11′MTT	[39] 17′JSSC	[40] 09′ISSCC	[41] 20′MWCL
Technology	65-nm CMOS	65-nm CMOS	65-nm CMOS	0.13-μm CMOS	0.13-μm CMOS	0.13-μm CMOS	90-nm CMOS
Self-oscillation frequency (GHz)	14.08	17.5	N/A	5.9	25.9	N/A	9.7
Input signal power (dBm)	0	0	0	0	0	0	−5
Division ratio	2	2	4	4	2	2	6
Input frequency range (GHz)	18.8–33.8	31.7–39.3	58.5–72.9	13.5–30.5	35–44 41–59.5	35.6–39.3	54.5–60.1
Locking range (%)	57	21.4	21.9	77.3	53 *	9.9	9.8
Supply Voltage (V)	1	1	0.6	1.4	1.15	1	N/A
Power consumption of core (mW)	2.26	2.5	2.2	7.3	3.8	3.12	5.6
Phase noise (dBc/Hz @1 MHz)	−129.81 (14 GHz)	−102 (N/A)	−126.74 (N/A)	−137.4 (6 GHz)	−124 (24 GHz)	−133.7 (N/A)	−140 (9.7 GHz)
FOM$_1$ (GHz/mW)	6.64	3.04	6.54	2.33	6.45	1.19	1
FOM$_2$ (GHz/mW)	13.28	6.08	26.16	9.32	12.9	2.38	6
Chip size (mm^2)	0.75 × 0.45	0.6 × 0.75	0.16 × 0.26	0.52 × 0.64	1 × 0.9	0.13 × 0.18 **	0.83 × 0.61

*: Total locking range (low band + high band), **: Only core size.

5. Conclusions

This paper presents the wide locking range and low-power divide-by-two CR-ILFD. The fourth-order resonator is applied to extend the narrow operating range of the ILFD. In addition, the CR core decreases the power consumption. The input frequency locking range is from 18.8 to 33.8 GHz (57%) at an injection power of 0 dBm. The full-span spectrums at the maximum or minimum frequency are presented. The power difference between the output and harmonic signals is approximately 10 dB or more over the entire locking range. The proposed CR-ILFD dissipates 2.26 mW from a 1 V supply voltage and the die size is 0.75 mm × 0.45 mm. This CR-ILFD is implemented in a 65 nm CMOS technology.

Author Contributions: Conceptualization, K.-I.O. and D.B.; methodology, K.-I.O.; software, K.-I.O.; validation, K.-I.O., G.-H.K. and D.B.; formal analysis, K.-I.O.; investigation, K.-I.O. and G.-H.K.; resources, K.-I.O.; data curation, K.-I.O.; writing—original draft preparation, K.-I.O., J.-G.K. and D.B.; writing—review and editing, K.-I.O. and D.B.; visualization, K.-I.O. and D.B.; supervision, K.-I.O. and D.B.; project administration, K.-I.O., J.-G.K. and D.B.; funding acquisition, D.B. All authors have read and agreed to the published version of the manuscript.

Funding: This work was supported by the Institute for Information & Communications Technology Planning & Evaluation (IITP) grant funded by the Korea government (MSIT) (No. 2019-0-00138, Development of Intelligent Radar Platform Technology for Smart Environments) and by the Chung-Ang University Research Scholarship Grants in 2019.

Institutional Review Board Statement: Not applicable.

Informed Consent Statement: Not applicable.

Data Availability Statement: No new data were created in this study. Data sharing is not applicable to this article.

Acknowledgments: The authors would like to thank all authors of previous papers for approving the use of their published research results in this paper.

Conflicts of Interest: The authors declare no conflict of interest.

References

1. Kim, K.-R.; Kim, J.-H. Wideband Waveform Generation Using MDDS and Phase Compensation for X-Band SAR. *Sensors* **2020**, *12*, 1431. [CrossRef]
2. Song, J.-H.; Lee, K.-W.; Lee, W.-K.; Jung, C.-H. High Resolution Full-Aperture ISAR Processing through Modified Doppler History Based Motion Compensation. *Sensors* **2017**, *12*, 1234. [CrossRef] [PubMed]
3. Park, J.; Ryu, H.; Ha, K.-W.; Kim, J.-G.; Baek, D. 76–81-GHz CMOS Transmitter with a Phase-Locked-Loop-Based Multichirp Modulator for Automotive Radar. *IEEE Trans. Microw. Theory Tech.* **2015**, *63*, 1399–1408. [CrossRef]
4. Ek, S.; Pahlsson, T.; Elgaard, C.; Carlsson, A.; Axholt, A.; Stenman, A.K.; Sundstrom, L.; Sjoland, H. A 28-nm FD-SOI 115-fs Jitter PLL-Based LO System for 24-30-GHz Sliding-IF 5G Transceivers. *IEEE J. Solid-State Circuits* **2018**, *53*, 1988–2000. [CrossRef]
5. Wang, H.-N.; Huang, Y.-W.; Chung, S.-J. Spatial Diversity 24-GHz FMCW Radar with Ground Effect Compensation for Automotive Applications. *IEEE Trans. Veh. Technol.* **2017**, *66*, 965–973. [CrossRef]
6. Lee, J.-Y.; Kim, G.S.; Ko, G.-H.; Oh, K.-I.; Park, J.G.; Baek, D. Low Phase Noise and Wide-Range Class-C VCO Using Auto-Adaptive Bias Technique. *Electronics* **2020**, *9*, 1290. [CrossRef]
7. Kang, C.-W.; Moon, H.; Yang, J.-R. Switched-Biasing Techniques for CMOS Voltage-Controlled Oscillator. *Sensors* **2021**, *21*, 316. [CrossRef] [PubMed]
8. Oh, K.-I.; Kim, G.S.; Park, J.-G.; Ko, G.-H.; Baek, D. Fourth-order Resonator based Current Reuse Injection-Locking Frequency Divider with Peaking Inductor. In Proceedings of the 2019 Asia-Pacific Microwave Conference (APMC), Singapore, 10–13 December 2019; pp. 1421–1423.
9. Zhao, X.; Chen, Y.; Mak, P.-I.; Martins, R. A 0.0018-mm2 153% Locking-Range CML-Based Divider-by-2 With Tunable Self-Resonant Frequency Using an Auxiliary Negative-gm Cell. *IEEE Trans. Circuits Syst. I Reg. Pap.* **2019**, *66*, 3330–3339. [CrossRef]
10. Chen, Y.; Yang, Z.; Zhao, X.; Huang, Y.; Mak, P.-I.; Martins, R.P. A $6.5 \times 7 \mu m^2$ 0.98-to-1.5 mW Nonself-Oscillation-Mode Frequency Divider-by-2 Achieving a Single-Band Untuned Locking Range of 166.6% (4–44 GHz). *IEEE J. Solid-State Circuits Lett.* **2019**, *2*, 37–40. [CrossRef]
11. Ghilioni, A.; Mazzanti, A.; Svelto, F. Analysis and design of mm-Wave frequency dividers based on dynamic latches with load modulation. *IEEE J. Solid-State Circuits* **2013**, *48*, 1842–1850. [CrossRef]
12. Hussein, A.I.; Paramesh, J. Design and Self-Calibration Techniques for Inductor-Less Millimeter-Wave Frequency Dividers. *IEEE J. Solid-State Circuits* **2017**, *52*, 1521–1541. [CrossRef]
13. Lin, Y.-H.; Wang, H. A 35.7-64.2 GHz low power Miller Divider with Weak Inversion Mixer in 65 nm CMOS. *IEEE Microw. Wireless Compon. Lett.* **2016**, *26*, 948–950. [CrossRef]
14. Lin, Y.-S.; Huang, W.-H.; Lu, C.-L.; Wang, Y.-H. Wide-Locking-Range Multi-Phase-Outputs Regenerative Frequency Dividers Using Even-Harmonic Mixers and CML Loop Dividers. *IEEE Trans. Microw. Theory Tech.* **2014**, *62*, 3065–3075. [CrossRef]
15. Kuo, Y.-H.; Tsai, J.-H.; Huang, T.-W.; Wang, H. Design and Analysis of Digital-Assisted Bandwidth-Enhanced Miller Divider in 0.18- m CMOS Process. *IEEE Trans. Microw. Theory Tech.* **2012**, *60*, 3769–3777. [CrossRef]
16. Chen, Y.-T.; Li, S.-W.; Huang, T.-H.; Chuang, H.-R. V-Band CMOS Direct Injection-Locked Frequency Divider Using Forward Body Bias Technology. *IEEE Microw. Wireless Compon. Lett.* **2010**, *20*, 396–398. [CrossRef]
17. Cheng, Y.-H.; Tsai, J.-H.; Huang, T.-H. Design of a 90.9% Locking Range Injection-Locked Frequency Divider with Device Ratio Optimization in 90-nm CMOS. *IEEE Trans. Microw. Theory Tech.* **2017**, *65*, 187–197. [CrossRef]
18. Jang, S.-L.; Lai, W.-C.; Ciou, Y.-L.; Hou, J.C. Divide-by-2 Injection-Locked Frequency Dividers Using the Electric-Field Coupling Dual-Resonance Resonator. *IEEE Trans. Microw. Theory Tech.* **2020**, *68*, 844–853. [CrossRef]
19. Zou, Q.; Ma, K.; Yeo, K.S. A V-Band Wide Locking Range Divide-by-4 Injection-Locked Frequency Divider. *IEEE Microw. Wireless Compon. Lett.* **2018**, *28*, 1020–1022. [CrossRef]
20. Kim, J.; Lee, S.; Choi, D.-H. Injection-Locked Frequency Divider Topology and Design Techniques for Wide Locking-Range and High-Order Division. *IEEE Access* **2016**, *5*, 4410–4417. [CrossRef]
21. Zhang, J.; Cheng, Y.; Zhao, C.; Wu, Y.; Kang, K. Analysis and Design of Ultra-Wideband mm-Wave Injection-Locked Frequency Dividers Using Transformer-Based High-Order Resonators. *IEEE J. Solid-State Circuits* **2018**, *53*, 2177–2189. [CrossRef]
22. Chao, Y.; Luong, H. Analysis and Design of Wide-Band Millimeter-Wave Transformer-Based VCO and ILFDs. *IEEE Trans. Circuits Syst. I Reg. Pap.* **2016**, *63*, 1416–1425. [CrossRef]
23. Ha, K.-W.; Ryu, H.; Lee, J.-H.; K, J.-G.; Baek, D. Gm-Boosted Complementary Current-Reuse Colpitts VCO With Low Power and Low Phase Noise. *IEEE Microw. Wireless Compon. Lett.* **2014**, *24*, 418–420. [CrossRef]
24. Ha, K.-W.; Ryu, H.; Park, J.; K, J.-G.; Baek, D. Transformer-Based Current-Reuse Armstrong and Armstrong-Colpitts VCOs. *IEEE Trans. Circuits Syst. II Exp. Briefs* **2014**, *61*, 676–680. [CrossRef]
25. Ryu, H.; Ha, K.W.; Baek, D. Low-power quadrature voltage-controlled oscillator using current-reuse and transformer-based Armstrong topologies. *Electron. Lett.* **2016**, *52*, 462–464. [CrossRef]
26. Yun, S.J.; Shin, S.-B.; Choi, H.-C.; Lee, S.-G. A 1mW Current-Reuse CMOS Differential LC-VCO with Low Phase Noise. In Proceedings of the 2005 IEEE International Solid-State Circuits Conference (ISSCC) Digest of Technical Papers, San Francisco, CA, USA, 11–15 February 2005; pp. 540–541.
27. Ha, K.-W.; Sung, E.-T.; Baek, D. Low-power transformer-based current-reuse injection-locked frequency divider. *Electron. Lett.* **2015**, *51*, 161–162. [CrossRef]

28. Lee, I.-T.; Tsai, K.-H.; Liu, S.-I. A 104- to 112.8-GHz CMOS Injection-Locked Frequency Divider. *IEEE Trans. Circuits Syst. II Exp. Briefs* **2009**, *56*, 555–559.
29. Lin, B.-Y.; Liu, S.-I. Analysis and Design of D-Band Injection-Locked Frequency Dividers. *IEEE J. Solid-State Circuits* **2011**, *46*, 1250–1264. [CrossRef]
30. Chine, K.-H.; Chen, J.-Y.; Chiou, H.-K. Designs of K-Band Divide-by-2 and Divide-by-3 Injection-Locked Frequency Divider with Darlington Topology. *IEEE Trans. Microw. Theory Tech.* **2015**, *63*, 2877–2888. [CrossRef]
31. Jang, S.-L.; Luo, J.-C.; Chang, C.-W.; Lee, C.-F.; Huang, J.-F. LC-Tank Colpitts Injection-Locked Frequency Divider with Even and Odd Modulo. *IEEE Microw. Wireless Compon. Lett.* **2009**, *19*, 113–115. [CrossRef]
32. Jang, S.-L.; Huang, S.-H.; Lee, C.-F.; Juang, M.-H. LC-Tank Colpitts Injection-Locked Frequency Divider with Record Locking Range. *IEEE Microw. Wireless Compon. Lett.* **2008**, *18*, 560–562. [CrossRef]
33. Mahalingam, N.; Ma, K.; Yeo, K.-S.; Lim, W.-M. Coupled Dual LC Tanks Based ILFD With Low Injection Power and Compact Size. *IEEE Microw. Wireless Compon. Lett.* **2014**, *24*, 105–107. [CrossRef]
34. Wu, J.-W.; Tu, C.-H.; Chen, S.-W.; Chen, C.-C.; Hsiao, H.-F.; Chang, D.C. Low Power Consumption and Wide Locking Range Triple-Injection-Locked Frequency Divider by Two. In Proceedings of the 2015 Asia-Pacific Microwave Conference (APMC), Nanjing, China, 6–9 December 2015; pp. 1–3.
35. Jang, S.-L.; Jina, S.-J.; Hsue, C.-W. Wideband Divide-by-4 Injection-Locked Frequency Divider Using Harmonic Mixer. *IEEE Microw. Wireless Compon. Lett.* **2017**, *27*, 924–926. [CrossRef]
36. Mahalingam, N.; Ma, K.; Yeo, K.S.; Lim, W.M. Modified Inductive Peaking Direct Injection ILFD With Multi-Coupled Coils. *IEEE Microw. Wireless Compon. Lett.* **2015**, *25*, 379–381. [CrossRef]
37. Wu, L.; Luong, H. Analysis and Design of a 0.6 V 2.2 mW 58.5-to-72.9 GHz Divide-by-4 Injection-Locked Frequency Divider with Harmonic Boosting. *IEEE Trans. Circuits Syst. I Reg. Pap.* **2013**, *60*, 2001–2008. [CrossRef]
38. Kuo, Y.-H.; Tsai, J.-H.; Chang, H.-Y.; Tsai, T.-W. Design and Analysis of a 77.3% Locking-Range Divide-by-4 Frequency Divider. *IEEE Trans. Microw. Theory Tech.* **2011**, *59*, 2477–2485. [CrossRef]
39. Imani, A.; Hashemi, H. Distributed Injection-Locked Frequency Dividers. *IEEE J. Solid-State Circuits* **2017**, *52*, 2083–2093. [CrossRef]
40. Chen, H.-K.; Chen, H.H.; Chang, D.-C.; Juang, Y.-Z.; Yang, Y.-C.; Lu, S.-S. A mm-Wave CMOS Multimode Frequency Divider. In Proceedings of the 2009 IEEE International Solid-State Circuits Conference (ISSCC) Digest of Technical Papers, San Francisco, CA, USA, 8–12 February 2009; pp. 280–282.
41. Chang, H.-Y.; Chen, W.-C.; Yeh, H.-N.; Shen, I.Y.-E. A V-Band CMOS Low-DC-Power Wide-Locking-Range Divide-by-6 Injection-Locked Frequency Divider Using Transformer Coupling. *IEEE Microw. Wireless Compon. Lett.* **2020**, *30*, 593–596. [CrossRef]

Communication

A New Current-Shaping Technique Based on a Feedback Injection Mechanism to Reduce VCO Phase Noise

Francisco Javier del Pino Suárez [†] and Sunil Lalchand Khemchandani [*,†]

Departamento de Ingeniería Electrónica y Automática, Institute for Applied Microelectronics (IUMA), Universidad de Las Palmas de Gran Canaria, E-35017 Las Palmas de Gran Canaria, Spain; jpino@iuma.ulpgc.es
* Correspondence: sunil.lalchand@ulpgc.es
† These authors contributed equally to this work.

Abstract: Inductor-capacitor voltage controlled oscillators (LC-VCOs) are the most common type of oscillator used in sensors systems, such as transceivers for wireless sensor networks (WSNs), VCO-based reading circuits, VCO-based radar sensors, etc. This work presents a technique to reduce the LC-VCOs phase noise using a new current-shaping method based on a feedback injection mechanism with only two additional transistors. This technique consists of keeping the negative resistance seen from LC tank constant throughout the oscillation cycle, achieving a significant phase noise reduction with a very low area increase. To test this method an LC-VCO was designed, fabricated and measured on a wafer using 90 nm CMOS technology with 1.2 V supply voltage. The oscillator outputs were buffered using source followers to provide additional isolation from load variations and to boost the output power. The tank was tuned to 1.8 GHz, comprising two 1.15 nH with 1.5 turns inductors with a quality factor (Q) of 14, a 3.27 pF metal-oxide-metal capacitor, and two varactors. The measured phase noise was -112 dBc/Hz at 1 MHz offset. Including the pads, the chip area is $750 \times 850 \ \mu m^2$.

Keywords: LC-VCO; CMOS; phase noise; current-shaping; 90 nm; current tail; varactor; LC tank; on-wafer

Citation: del Pino Suárez, F.J.; Khemchandani, S.L. A New Current-Shaping Technique Based on a Feedback Injection Mechanism to Reduce VCO Phase Noise. *Sensors* **2021**, *21*, 6583. https://doi.org/10.3390/s21196583

Academic Editors: Jong-Ryul Yang and Seong-Tae Han

Received: 1 September 2021
Accepted: 28 September 2021
Published: 1 October 2021

1. Introduction

Voltage-controlled oscillators (VCOs) are widely used in the design of sensor systems. VCOs are generally found in transceivers for ultra low-power wireless sensor networks (WSNs) where, in conjunction with the phase locked loop (PLL), are used for frequency synthesis, fast switching circuits, and clock recovery [1–10]. VCOs are also a fundamental part of VCO-based reading circuits where the sensor core output voltage is applied to the VCO tuning voltage node, achieving high sensitivity and high signal-to-noise ratio compared to amplifier-based reading circuits [11–14]. Also, distance, speed and other parameters can be remotely measured in real time using VCO-based radar sensors that monitor the electromagnetic wave shift between transmitted and received signals [15–17]. The purity of the VCO output signal greatly influences the operation of these systems and, for this reason, the main considerations when designing a VCO are low phase noise, minimal chip area, low power dissipation, and high operating frequency. Today's nanoscale complementary metal-oxide semiconductor (CMOS) technology can meet most of these requirements. However, reducing phase noise is still a major issue, mainly due to the poor performance of CMOS process in terms of flicker noise [10]. The major approach that has been used to reduce VCO flicker noise is to apply biasing techniques to both the VCO core transistors and the current source transistors needed to supply the DC bias current of the core transistors. In [18], a review of techniques for reducing CMOS VCO phase noise caused by flicker noise is presented. This study focuses on current source transistors biasing techniques and concludes that current-shaping techniques can significantly reduce its flicker noise contribution to the output phase noise.

41

In this paper a new current-shaping technique is proposed to reduce the VCO's phase noise. The proposed technique is based on a feedback injection mechanism and only uses two additional transistors. Using this approach, a significant phase noise reduction is achieved with a very low area increase. Section 2 introduces the techniques for the reduction of the phase noise of CMOS based VCOs and describes the advantages of the current-shaping techniques. Section 3 presents the proposed topology and analysis, followed by the experimental results in Section 4. Finally, Section 5 concludes the paper.

2. Current-Shaping Biasing Techniques

One of the most used topologies for the implementation of VCO circuits is the inductor-capacitor voltage controlled oscillator (LC-VCO) since they show less phase noise although they occupy a high area due to the presence of inductors and dissipate more power. Figure 1 shows the conventional structure of an LC-VCO. The bulk of the NMOS transistors were connected to lowest potential which is ground. The close-in phase noise behaviour at an offset Δf from the carrier frequency f_0 is given by Leeson's model [19].

$$L(\Delta f) = \frac{1}{2} \frac{K\,T\,F}{P_{sig}} \left(1 + \frac{f_c}{\Delta f}\right) \left(1 + \frac{f_0}{2\,Q\,\Delta f}\right)^2, \tag{1}$$

where K is Boltzmann's constant, T is the absolute temperature, F is the excess noise factor, P_{sig} is the signal power, Q is the resonator loaded quality factor, and f_c is the flicker noise corner where flicker noise and thermal noise are equal. This equation leads to the typical plot of phase noise versus offset frequency of Figure 2 and it also offers design insight on how to minimise the overall phase noise. It is well known that a lower excess noise factor (F), a larger amplitude of oscillation (P_{sig}), or a better tank quality factor (Q) results in an improved phase noise.

The previous analysis is based on a linear time invariant analysis of the oscillator. However, a more detailed analysis based on transient simulations indicates that the tail current bias noise can contribute strongly to the total phase noise [20]. The reason behind this phenomenon is that the switching transistors (M1 and M2) behave like an up-conversion mixer and convert the flicker noise of the tail current into AM noise at the output of the VCO which is subsequently converted into PM noise by the non-linear varactor. Also, the same mixing mechanism convert the tail current noise at the harmonics of the frequency of oscillation (ω_0) directly to PM noise at the output through the indirect FM phenomenon [10,20].

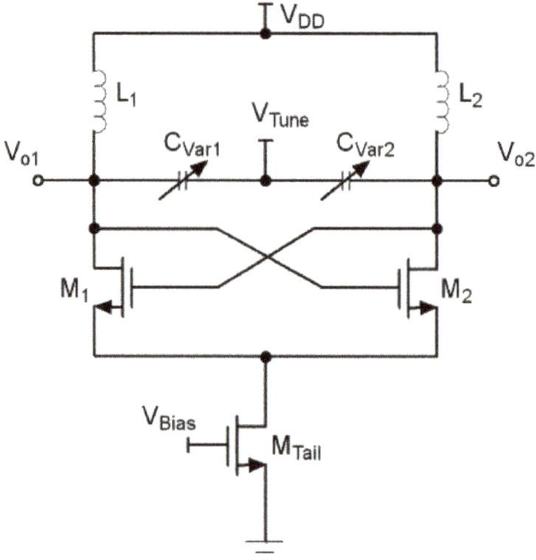

Figure 1. Conventional LC-VCO.

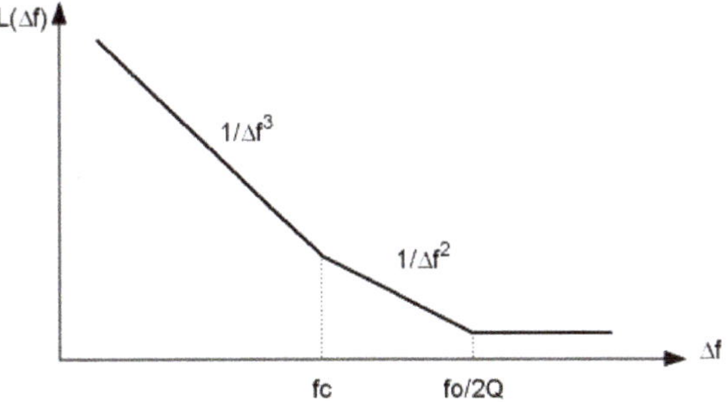

Figure 2. Phase noise vs. frequency.

A possible solution to remove the harmonics of the tail current source is to filter them out by a capacitor, as shown in Figure 3 [21]. This technique is known as tail current-shaping and consist of reducing the tail current at the moments when the oscillator is most sensitive to noise, that is, when its effective impulse sensitivity function (ISF) is higher. The more symmetrical the shape of the tail current, the fewer harmonics it will have and therefore the lower the phase noise of the VCO.

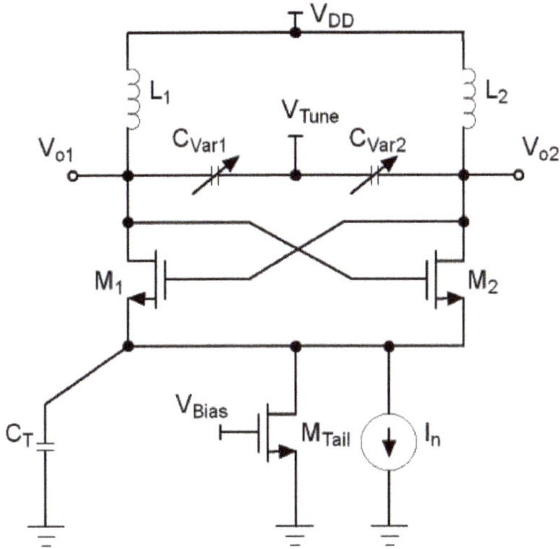

Figure 3. LC-VCO with tail current-shaping.

The contribution of the tail current bias noise to the total phase noise is further aggravated if during the oscillation M1 and M2 enter the deep triode region. If this occurs, then two effects raise the phase noise. First, the on-resistance of M1 and M2 degrades the Q of the tank and second, the impulse response from the noise of both transistors contributes substantially to the output phase noise.

To avoid operation in the triode region but at the same time have large output swings, capacitive coupling can be inserted in the loop as shown in Figure 4 [22]. This topology is known as Class-C oscillator. In this topology, a bias voltage (V_B) is chosen so that M1 and M2 operate in the active region and, at the same time, allow a sufficiently high voltage swing at the drains to improve the phase noise performance. However, the peak output swing is limited by the tail capacitor and therefore, once a maximum is reached, no further improvement in phase noise can be achieved.

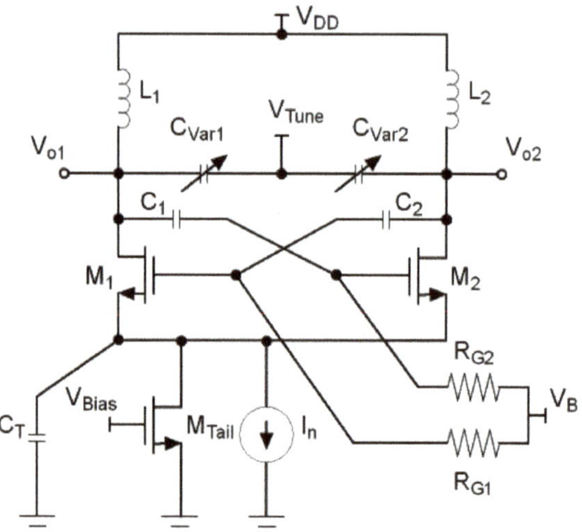

Figure 4. Class-C LC-VCO.

Another approach is to allow the switching transistors to enter the triode region but eliminating the effect of the tail capacitance at $2\omega_0$ introducing an inductor in series with the tail node, as shown in Figure 5 [23]. The value of the inductor is chosen such that it resonates with the parasitic capacitance at the tail node at the second harmonic. With this topology larger swings can be achieved at the cost of a larger area due to the use of an additional inductor.

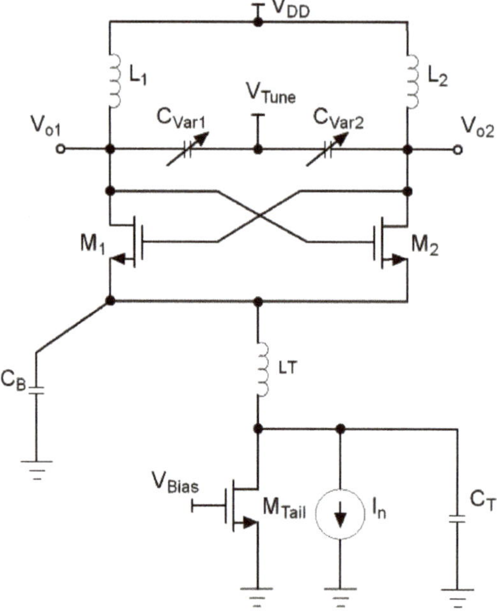

Figure 5. Filtering technique to lower LC-VCO phase noise.

Noise cancellation topologies such as presented in [20,24,25] have also been employed to cancel the tail current noise component. However, they are hardly effective at high frequencies because they are able to cancel only the contribution of flicker noise, leaving the thermal noise unaffected.

Many current-shaping techniques have been proposed in recent years to improve the VCO phase noise performance. One approach is the use of a switched biasing technique where an external pulse is injected into the gate of the tail current transistor [21]. This method has proven to reduce the phase noise, but its drawback is that an external pulse signal is required for locking. To avoid this, some authors have proposed to shape the tail current using the oscillator's own output waveform as self-injection signal. A simple but high area cost method is to couple the oscillating signal VCO to the tail current source through a transformer [26]. Another approach is to couple the VCO output voltage directly to the gates of two tail transistors as shown in Figure 6 [27]. Although this method considerably reduces the area, the downside is that both, the AC and the DC parts of the output voltage are coupled to the tail transistors, resulting in high bias voltage of the tail current sources and, as a consequence, in high power consumption and flicker noise. One way to avoid this is to decouple the DC part from the AC part of the output voltage using a capacitor as shown in Figure 7 [28–31]. In this way, the DC part of the gate voltage of the tail transistors comes from an external source while the AC part comes from the VCO output. This allows to choose a DC voltage small enough so that the current supplied to the switching transistors is significantly reduced at the zero crossing points of the output, thereby reducing phase noise.

Feedback injection currents have also been proposed to improve the phase noise by modifying the triode region loading effect of the switch transistors and increasing the output transconductance. This method also causes a self-locking between the output voltage and currents at the zero crossing points, further reducing phase noise [32]. In this paper we propose a new current-shaping technique based on a feedback injection mechanism. The proposed topology significantly reduces phase noise by using only two additional transistors. In the next section we describe the proposed technique.

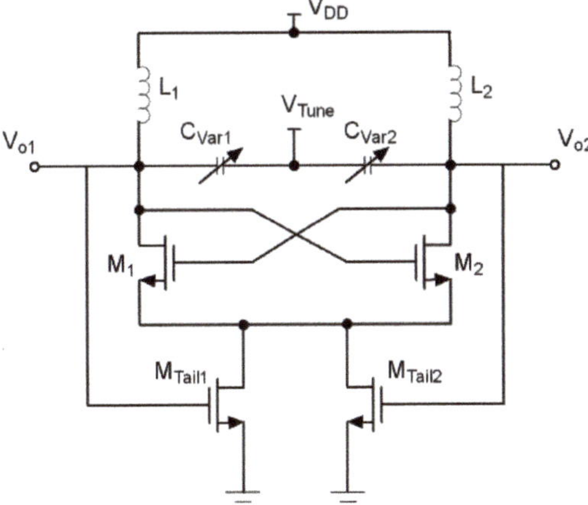

Figure 6. Tail current-shaping by coupling the VCO output voltage directly to the gates of two tail transistors.

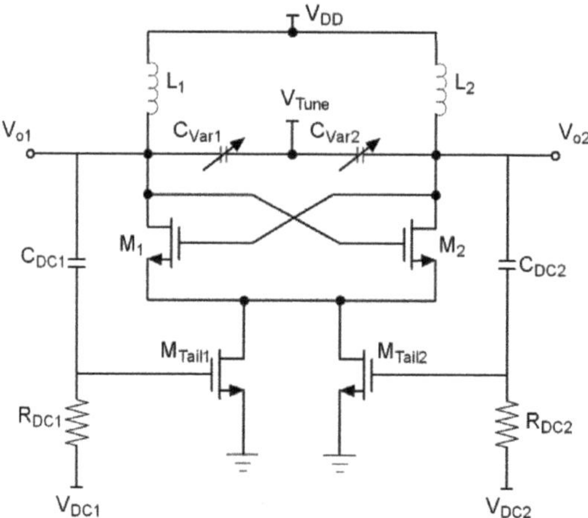

Figure 7. Tail current-shaping by coupling the VCO output voltage to the gates of two tail transistors decoupling the DC part from the AC part using a capacitor.

3. Proposed Topology

The schematic of the proposed VCO is shown in Figure 8. The VCO core uses a cross-coupled transistor pair, M1 and M2, to build up the negative resistance. To ensure the loading effect and to improve the current feedback, i_{fb} and $-i_{fb}$, the network composed by M3 and M4 is included.

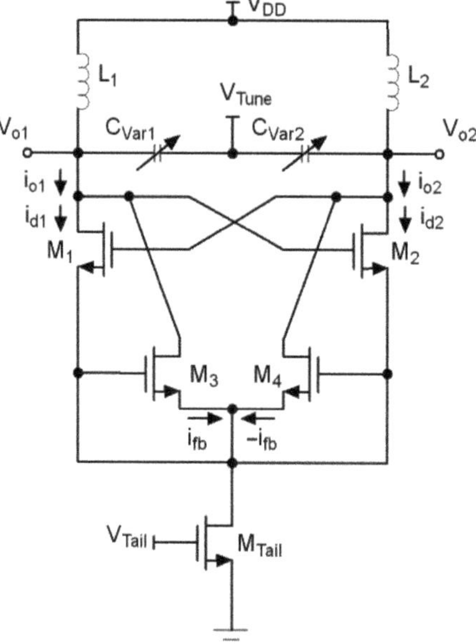

Figure 8. Proposed feedback injection mechanism to reduce the VCO's phase noise.

In a conventional VCO, V_{o1} and V_{o2} voltages decrease and increase respectively during a half cycle of oscillation. If the cut-off regions are ignored, M1 and M2 swing between triode and saturation regions. For this reason, the drain-source resistances are not the same in both regions, being lower in triode. This increases the loading effect on the LC tank in this region, thus degrading the phase noise of the VCO.

In the proposed circuit, when V_{o1} decreases and V_{o2} increases, a feedback current i_{fb} from node V_{o2} is injected into node V_{o1} through M4 and M3 transistors. This increases i_{d1} and pulls M1 back from triode towards saturation region. The impedance seen from the LC tank towards the V_{o1} node rises due to the virtual magnification of the drain-source resistance of M1. Then, the negative resistance seen from LC tank remains with the same value during the full oscillation cycle. The same explanation is applied to M2.

Figures 9 and 10 show the simulated output waveforms V_{o1} and V_{o2} and the tail current (I_{tail}) of a conventional LC-VCO and our proposal. The transistor models used are typical BSIM3 model that loosely model a 0.25 μm CMOS process. They are based on measured data from MOSIS for a 0.25 μm process and have been modified by to create the 3.3 V devices and 0.35 μm gate lengths using a thicker oxide. The models include noise and use the Advanced Compact MOSFET (ACM) model. The simulation shows that in the conventional case I_{tail} is not symmetric while our proposed VCO presents a much more symmetric I_{tail}. Figure 11 compares the phase noise of our proposed VCO with the conventional VCO. The phase noise of our proposed VCO is -105.4 dBc/Hz at 100 kHz offset frequency, which is 7.5 dB lower than the simulated conventional VCO. Due to the parasitics introduced by M3 and M4 to the LC tank, the proposed solution lowers the output frequency. This effect is minor, but must be taken into account when using this topology.

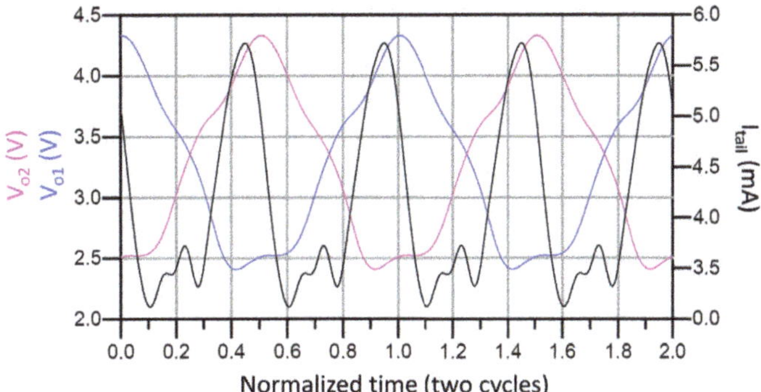

Figure 9. V_{o1} and V_{o2} and I_{tail} of the conventional VCO.

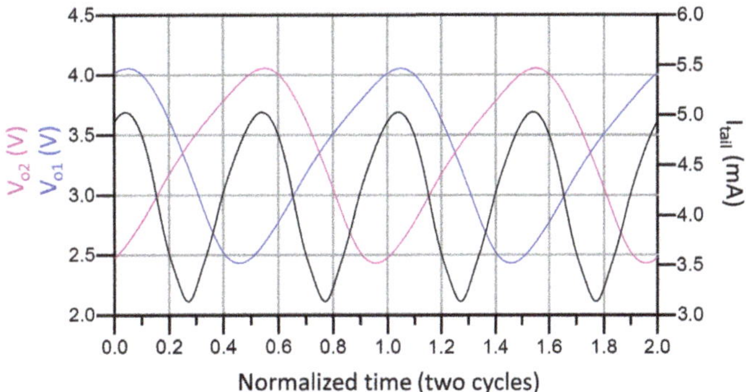

Figure 10. V_{o1} and V_{o2} and I_{tail} of the proposed VCO.

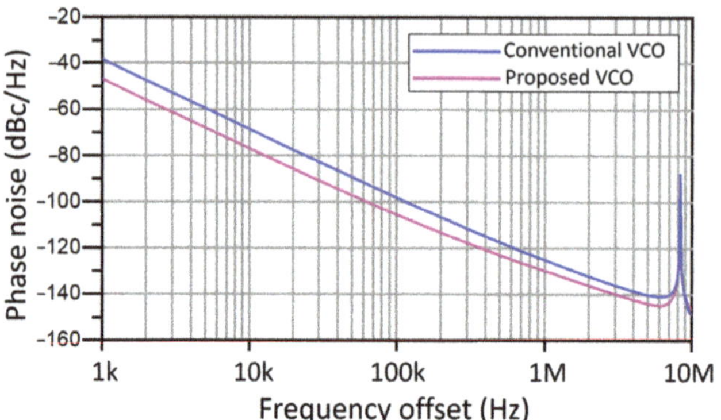

Figure 11. Phase noise of the conventional and proposed VCO.

4. Measurement Results

To test the proposed topology, a prototype chip was designed using UMC 90 nm CMOS process. Figures 12 and 13 shows the circuit simplified schematic and microphotograph, respectively. The oscillator outputs are buffered using CMOS source followers to provide additional isolation from load variations and to boost the output power. The tank was tuned to 1.8 GHz, comprising two 1.15 nH with 1.5 turns inductors with a Q of 14, a 3.27 pF metal-oxide-metal (MOM) capacitor and two varactors. A voltage applied to the V_{Tune} pin, which is connected to varactors, controls the VCO oscillation frequency. The total area occupied by the circuit is $750 \times 850\ \mu m^2$ including the pads for on wafer measurement. Table 1 summarizes the value of the components of the VCO.

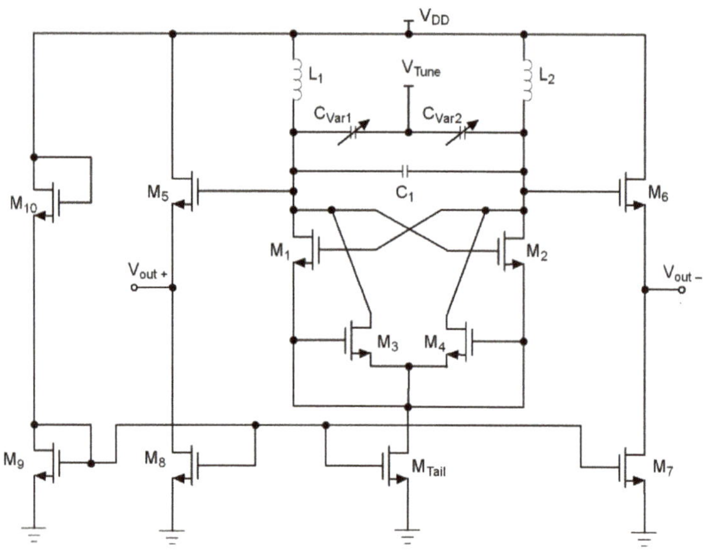

Figure 12. Simplified schematic of the fabricated VCO.

Figure 14 shows the measured frequency spectrum of the proposed VCO when $V_{Tune} = 0$, the output power is -12 dBm at 1.83 GHz oscillating frequency. The measured loss of the combination of the probe and cable is 1.1 dB, so the output power is -10.87 dBm. As shown in Figure 15, the output power keeps almost constant while the oscillation frequency can be tuned from 1.72 GHz to 1.83 GHz as the control voltage ranges from 1.2 to 0 V. The measured phase noise is shown in Figure 16. Table 2 compares the measured and simulated results indicating a good agreement between simulation and measurement. The VCO prototype core consumes 3.3 mA at 1.2 V supply and the total power consumption of the VCO, including the output buffers and bias, is 15.84 mW.

Table 1. VCO componentes values.

Component	Value
M_1 & M_2	W_{finger} = 1 µm, L = 200 nm, Multiplicity = 20
M_3 & M_4	W_{finger} = 8 µm, L = 360 nm, Multiplicity = 30
M_5 & M_6	W_{finger} = 530 nm, L = 100 nm, Multiplicity = 8
M_7 & M_8	W_{finger} = 530 nm, L = 100 nm, Multiplicity = 8
M_9	W_{finger} = 600 nm, L = 250 nm, Multiplicity = 10
M_{10}	W_{finger} = 500 nm, L = 100 nm, Multiplicity = 14
C_{VAR1} & C_{VAR1}	C_{MAX} = 3.824 pF
L_1 & L_2	L = 1.15 nH, Q = 14@2.2 GHz, 1.5 turns
C_1	C = 1.364 pF

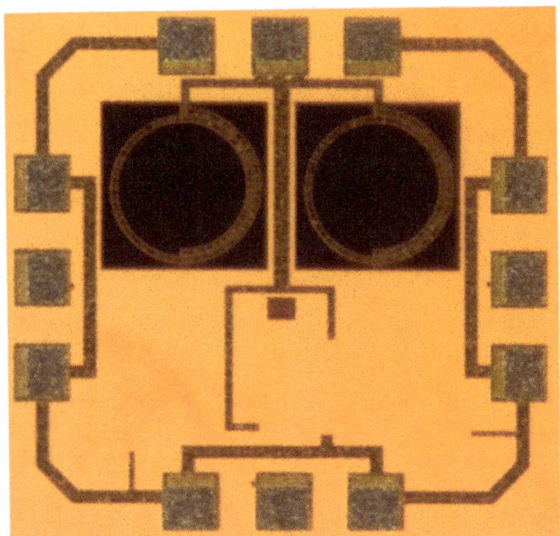

Figure 13. Microphotograph of the fabricated VCO.

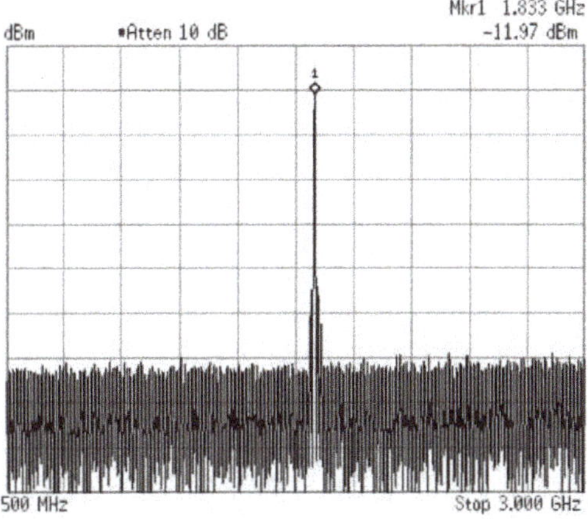

Figure 14. VCO Output Spectrum.

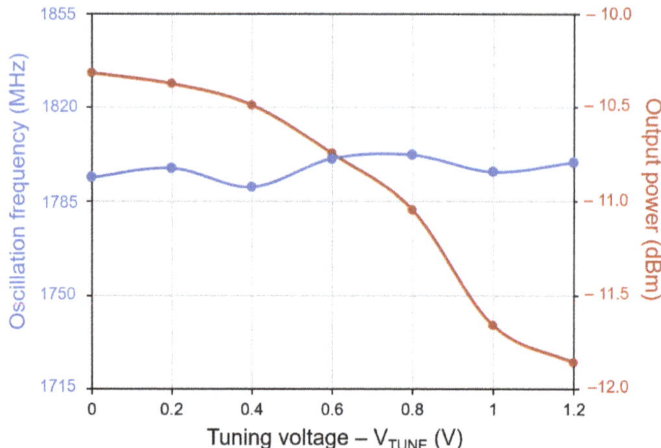

Figure 15. Measured frequency and output power vs. tuning voltage.

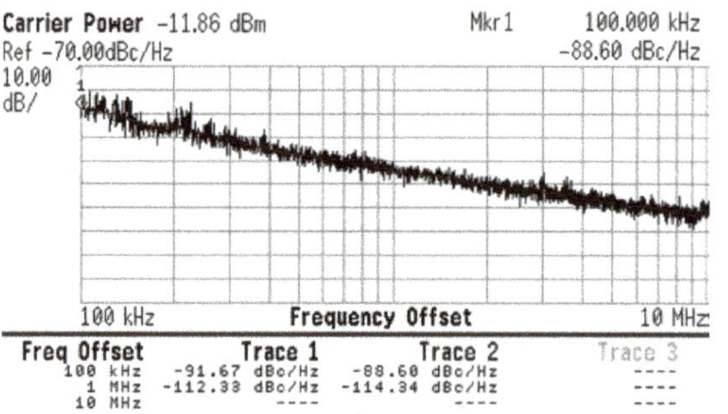

Figure 16. VCO measured phase noise.

Table 2. VCO simulated and measured phase noise. Average values for 1793 and 1833 MHz.

Frequency Offset	Simulated Phase Noise	Measured Phase Noise
100 kHz	−85 dBc/Hz	−86.6 dBc/Hz
1 MHz	−111.5 dBc/Hz	−112.2 dBc/Hz
5 MHz	−130 dBc/Hz	−125.5 dBc/Hz

A brief overview of similar works available in the literature is given in Table 3. The performances of the VCOs oscillating at different frequencies are compared using the typical figure-of-merit (*FoM*) [33]:

$$FoM[\text{dBc}] = L(\Delta f) - 20\log\left(\frac{f_0}{\Delta f}\right) + 10\log\left(\frac{P_{DC}}{1\,\text{mW}}\right),\tag{2}$$

where $L(\Delta f)$ is the phase noise (PN) in dB at the frequency offset Δf, f_0 is the center oscillation frequency and P_{DC} is the power consumption. *FoM* is specified at frequency offset of 1 MHz. As seen in Table 3, CMOS VCOs exhibit better performance than their NMOS counterparts. This is because CMOS structures provide higher transconductance for a given bias current [34]. As can be derived from Table 3, the VCO proposed in this paper compares to reported NMOS VCOs. Its main disadvantage is the operating frequency and tuning range, but this is due to the fact that we have designed the VCO for a lower frequency and tuning range. If a capacitor bank and a higher operating frequency where used, a better *FoM* would be achieved. The presented VCO phase noise and power consumption are comparable with the reported self-biasing NMOS VCOs. This is accomplished by adding just two transistors, and no additional electronics is required to generate the biasing. This fact is reflected in the area, which is one of the lowest reported. The area could have been further reduced if a symmetrical inductor had been used in the tank instead of two conventional inductors.

Table 3. VCO performance comparison.

Ref.	Year	Bias Scheme	Process (nm)	VCO Type	Supply (V)	Freq. (GHz)	Tuning Range (%)	PN@1 MHz (dBc/Hz)	Power (mW)	Area (mm^2)	*FoM*@1 MHz (dBc/Hz)
[35]	2019	Self-biasing w. fixed DC	180	CMOS-LC	1.2	2.45	28.6	−120	1.73	0.938	−185
[29]	2019	Self-biasing w. fixed DC	180	CMOS-LC	0.8	1.4	18	−123	0.7	2.706	−187
[36]	2015	Self-biasing w. fixed DC	180	CMOS-LC	1.2	2.55	9.2	−123	3.2	0.332 *	−186
[37]	2015	Self-biasing w. fixed DC	130	CMOS-LC	1.4	2.4	1.7	−128	4.2	0.092 *	−190
[38]	2019	Self-biasing w. adaptative DC	180	CMOS-LC	1.2	4.55	4.3	−123	1.35	0.979	−195
[39]	2009	Self-biasing w. adaptative DC	130	NMOS-LC	0.6	4.85	10.2	−117	3.9	0.723	−185
[40]	2020	Self-biasing w. fixed DC	65	NMOS-LC	0.45	10.4	13.6	−115	2.7	0.660	−191
[41]	2011	Self-biasing w. fixed DC	65	NMOS-LC	1.2	20	17	−107	19.2	0.800	−180
This work	**2021**	**Feedback injection curr.**	**90**	**NMOS-LC**	**1.2**	**1.77**	**6.2**	**−112**	**3.96**	**0.638**	**−171**

* Dimensions excluding pads.

5. Conclusions

A new current-shaping technique to reduce VCO phase noise has been proposed. This method uses a feedback injection mechanism that only uses two transistors and reduces the phase noise as compared to the conventional designs, with a minimum penalty in area and power consumption. To test the proposed solution, simulations were performed and used for evaluation and comparison. Also, a prototype chip fabricated in a 90 nm CMOS process was used to verify the proposed solution. The oscillation frequency can be tuned from 1.72 GHz to 1.83 GHz with an output power of −12 dBm at 1.83 GHz. The power consumption of the core is 3.96 mW. The phase noise results, −112 dBc/Hz at 1 MHz offset, indicates a good agreement between simulation and measurement.

Author Contributions: F.J.d.P.S. and S.L.K. contributed equally to this work. All authors have read and agreed to the published version of the manuscript.

Funding: This work was funded in part by the Spanish Ministry of Science, Innovation and Universities (RTI2018-099189-B-C22) and by the Canary Agency for Research, Innovation and Information Society (ACIISI) of the Canary Islands Government by ProID2017010067 and TESIS2019010100 grants.

Conflicts of Interest: The authors declare no conflict of interest.

Abbreviations

The following abbreviations are used in this manuscript:

ACM	Advanced Compact MOSFET
CMOS	Complementary Metal-Oxide-Semiconductor
F	Noise Factor
FoM	Figure of Merit
LC-VCO	Inductor-Capacitor Voltage Controlled Oscillator
ISF	Impulse Sensitivity Function
MOM	Metal-Oxide-Metal
PLL	Phase Locked Loop
Q	Quality Factor
VCO	Voltage-Controlled Oscillator
WSNs	Wireless Sensor Networks

References

1. Kopta, V.; Enz, C.C. A 4-GHz Low-Power, Multi-User Approximate Zero-IF FM-UWB Transceiver for IoT. *IEEE J. Solid-State Circuits* **2019**, *54*, 2462–2474. [CrossRef]
2. Wang, K.; Otis, B.; Wang, Z. A 580-μW 2.4-GHz ZigBee Receiver Front End With Transformer Coupling Technique. *IEEE Microw. Wirel. Compon. Lett.* **2018**, *28*, 174–176. [CrossRef]

3. Li, D.; Liu, D.; Kang, C.; Zou, X. A Low Power Low Phase Noise Oscillator for MICS Transceivers. *Sensors* **2017**, *17*, 140. [CrossRef]
4. Selvakumar, A.; Zargham M.; Liscidini, A. Sub-mW Current Re-Use Receiver Front-End for Wireless Sensor Network Applications. *IEEE J. Solid-State Circuits* **2015**, *50*, 2965–2974. [CrossRef]
5. Wong, A. C. W. A 1V 5 mA Multimode IEEE 802.15.6/Bluetooth Low-Energy WBAN Transceiver for Biotelemetry Applications. *IEEE J. Solid-State Circuits* **2013**, *48*, 186–198. [CrossRef]
6. Thabet, H., Meillère, S., Masmoudi, M. A low power consumption CMOS differential-ring VCO for a wireless sensor. *Analog Integr. Circ. Sig. Process* **2012**, *73*, 731–740. [CrossRef]
7. Italia, A.; Ippolito, C. M.:Palmisano, G. A 1-mW 1.13–1.9 GHz CMOS LC-VCO Using Shunt-Connected Switched-Coupled Inductors. *IEEE Trans. Circuits Syst. I Reg. Pap.* **2012**, 59, 1145–1155. [CrossRef]
8. Zhang, X.; Mukhopadhyay, I.; Dokania, R.; Apsel, A. B. A 46-µW Self-Calibrated Gigahertz VCO for Low-Power Radios. *IEEE Trans. Circuits Syst. II Exp. Briefs* **2011**, *58*, 847–851. [CrossRef]
9. Khemchandani, S.L.; del Pino Suarez, J.; Diaz Ortega, R.; Hernandez, A. A fully integrated single core VCO with a wide tuning range for DVB-H. *Microw. Opt. Technol. Lett.* **2009**, *51*, 1338–1343. [CrossRef]
10. Razavi, B. *RF Microelectronics*, 2nd ed.; Prentice Hall Communications Engineering and Emerging Technologies Series; Prentice Hall: New York, NY, USA, 2011.
11. Quintero, A.; Cardes, F.; Perez, C.; Buffa, C.; Wiesbauer, A.; Hernandez, L. A VCO-Based CMOS Readout Circuit for Capacitive MEMS Microphones. *Sensors* **2019**, *19*, 4126. [CrossRef]
12. Elhadidy, O.; Elkholy, M.; Helmy, A. A.; Palermo, S.; Entesari, K. A CMOS Fractional-N PLL-Based Microwave Chemical Sensor With 1.5% Permittivity Accuracy. *IEEE Trans. Microw. Theory Tech.* **2013**, *61*, 3402–3416. [CrossRef]
13. Laemmle, B.; Schmalz, K.; Scheytt, J.C.; Weigel, R.; Kissinger, D. A 125-GHz Permittivity Sensor With Read-Out Circuit in a 250-nm SiGe BiCMOS Technology. *IEEE Trans. Microw. Theory Tech.* **2013**, *61*, 2185–2194. [CrossRef]
14. Jamal, F. I. Low-Power Miniature K-Band Sensors for Dielectric Characterization of Biomaterials. *IEEE Trans. Microw. Theory Tech.* **2017**, *65*, 1012–1023. [CrossRef]
15. Tseng, S.T.; Kao, Y.H.; Peng, C.C.; Liu, J.Y.; Chu, S.C.; Hong, G.F.; Chu, T.S. A 65-nm CMOS Low-Power Impulse Radar System for Human Respiratory Feature Extraction and Diagnosis on Respiratory Diseases. *IEEE Trans. Microw. Theory Tech.* **2016**, *64*, 1029–1041. [CrossRef]
16. Park, J.-H.; Yang, J.-R. Two-Tone Continuous-Wave Doppler Radar Based on Envelope Detection Method. *Microw. Opt. Technol. Lett.* **2020**, *62*, 3146–3150. [CrossRef]
17. Sim, J.Y.; Park, J.-H.; Yang, J.-R. Vital-Signs Detector Based on Frequency-Shift Keying Radar. *Sensors* **2020**, *20*, 5516. [CrossRef] [PubMed]
18. Kang, C.-W.; Moon, H.; Yang, J.-R. Switched-Biasing Techniques for CMOS Voltage-Controlled Oscillator. *Sensors* **2021**, *21*, 316. [CrossRef] [PubMed]
19. Lesson, D.B. A simple model of feedback oscillator noise spectrum. *Proc. IEEE* **1966**, *54*, 329–330. [CrossRef]
20. Heng, C.; Bansal, A.; Cheng, S. J. Techniques for improving CMOS VCO performance. In Proceedings of the 2009 IEEE International Symposium on Radio-Frequency Integration Technology (RFIT), Singapore, 9 January–11 December 2009; pp. 182–189. [CrossRef]
21. Soltanian, B.; Kinget, P. R. Tail Current-Shaping to Improve Phase Noise in LC Voltage-Controlled Oscillators. *IEEE J. Solid-State Circuits* **2006**, *41*, 1792–1802. [CrossRef]
22. Mazzanti, A.; Andreani, P. Class-C Harmonic CMOS VCOs, With a General Result on Phase Noise. *IEEE J. Solid-State Circuits* **2008**, *43*, 2716–2729. [CrossRef]
23. Hegazi, E.; Sjoland, H.; Abidi, A. A filtering technique to lower LC oscillator phase noise. *IEEE J. Solid-State Circuits* **2001**, *36*, 1921–1930. [CrossRef]
24. Lin, Y.; To, K. H.; Hamel, J. S.; Huang, W. M. A. Fully integrated 5 GHz CMOS VCOs with on chip low frequency feedback circuit for 1/f induced phase noise suppression. In Proceedings of the 28th European Solid-State Circuits Conference, Florence, Italy, 24–26 September 2002; Volume 36, pp. 551–554. [CrossRef]
25. Heng, C.; Bansal, A.; Zheng, Y. Design of 1.94-GHz CMOS Noise-Cancellation VCO. *IEEE Trans. Microw. Theory Tech.* **2011**, *59*, 368–374. [CrossRef]
26. Huang, T.; Tseng, Y. A 1 V 2.2 mW 7 GHz CMOS Quadrature VCO Using Current-Reuse and Cross-Coupled Transformer-Feedback Technology. *IEEE Microw. Wirel. Compon. Lett.* **2008**, *18*, 698–700. [CrossRef]
27. Boon, C. C.; Tseng, M. A.; Yeo, K. S.; Ma, J. G; Zhang, K. L. RF CMOS low-phase-noise LC oscillator through memory reduction tail transistor. *IEEE Trans. Circuits Syst. II Exp. Briefs* **2004**, *51*, 85–90. [CrossRef]
28. Jannesari, A.; Kamarei, M. Sinusoidal shaping of the ISF in LC oscillators. *Int. J. Circuit Theory Appl.* **2008**, *36*, 757–768. [CrossRef]
29. Hsieh, J.; Lin, K. A 0.7-mW LC Voltage-Controlled Oscillator Leveraging Switched Biasing Technique for Low Phase Noise. *IEEE Trans. Circuits Syst. II Express Briefs* **2019**, *66*, 1307–1310. [CrossRef]
30. Jafari, B.; Sheikhaei, S. Low phase noise LC-VCO with sinusoidal tail current-shaping using cascode current source. *AEU-Int. J. Electron. Commun.* **2018**, *83*, 1434–8411. [CrossRef]
31. Shen, I.; Huang, T.; Jou, C.F. A Low Phase Noise Quadrature VCO Using Symmetrical Tail Current-Shaping Technique. *IEEE Microw. Wirel. Compon. Lett.* **2010**, *20*, 399–401. [CrossRef]

32. Sadr, M.N.H.R.; Dousti, M.; Hajghassem, H. Phase noise and transconductance improvement in monolithic differential LC-VCOs. *Int. J. Electron.* **2011**, *98*, 1333–1345. [CrossRef]
33. Kinget, P. *Integrated GHz Voltage Controlled Oscillators*; Springer US: Boston, MA, USA, 1999; pp. 353–381. ISBN 978-1-4419-5101-4.
34. González Ramírez, D.; Lalchand Khemchandani, S.; del Pino, J.; Mayor-Duarte, D.; San Miguel-Montesdeoca, M.; Mateos-Angulo, S. Single event transients mitigation techniques for CMOS integrated VCOs. *Microelectron. J.* **2018**, *73*, 37–42. [CrossRef]
35. Shasidharan, P.; Ramiah, H.; Rajendran, J. A 2.2 to 2.9 GHz Complementary Class-C VCO with PMOS Tail-Current Source. Feedback Achieving—120 dBc/Hz Phase Noise at 1 MHz Offset. *IEEE Access* **2019**, *7*, 91325–91336. [CrossRef]
36. Chen, N.; Diao, S.; Huang, L.; Lin, F. Reduction of $1/f^3$ Phase Noise in LC Oscillator with Improved Self-Switched Biasing. *Analog Integr. Circuits Signal Process* **2015**, *84*, 19–27. [CrossRef]
37. Mostajeran, A.; Bakhtiar, M.S.; Afshari, E. A 2.4 GHz VCO with FOM of 190 dBc/Hz at 10 kHz-to −2 MHz Offset Frequencies in 0.13 μm CMOS Using an ISF Manipulation Technique. In Proceedings of the 2015 IEEE International Solid-State Circuits Conference—(ISSCC) Digest of Technical Papers, San Francisco, CA, USA, 22–26 February 2015; pp. 1–3. [CrossRef]
38. Narayanan, A.T.; Okada, K. A Pulse-Tail-Feedback LC-VCO with 700 Hz Flicker Noise Corner and −195 dBc FoM. *IEICE Trans. Electron.* **2019**, *E102*, 595–606. [CrossRef]
39. Min, B.-H.; Hyun, S.-B.; Yu, H.-K. Low Voltage CMOS LC-VCO with Switched Self-Biasing. *ETRI J.* **2009**, *31*, 755–764. [CrossRef]
40. Lee, H.S.; Kang, D.M.; Cho, S.J.; Byeon, C.W.; Park, C.S. Low-Power, Low-Phase-Noise Gm-Boosted 10-GHz VCO With Center-Tap Transformer and Stacked Transistor. *IEEE Trans. Circuits Syst. II Express Briefs* **2020**, *67*, 1710–1714. [CrossRef]
41. Musa, A.; Murakami, R.; Sato, T.; Chaivipas, W.; Okada, K.; Matsuzawa, A. A Low Phase Noise Quadrature Injection Locked Frequency Synthesizer for MM-Wave Applications. *IEEE J. Solid-State Circuits* **2011**, *46*, 2635–2649. [CrossRef]

Communication

Current Input Pixel-Level ADC with High SNR and Wide Dynamic Range for a Microbolometer

Jeongho Lee [1], Ilku Nam [2] and DooHyung Woo [1],*

[1] School of Information, Communications and Electronics Engineering, The Catholic University of Korea, Gyeonggi-do 14662, Korea; jesseily92@gmail.com
[2] Department of Electrical Engineering, Pusan National University, Busan 46241, Korea; nik@pusan.ac.kr
* Correspondence: cowpox@catholic.ac.kr; Tel.: +82-2-2164-4522

Abstract: A readout circuit incorporating a pixel-level analog-to-digital converter (ADC) is studied for two-dimensional medium wavelength infrared microbolometer arrays. The signal-to-noise ratio (SNR) and charge handling capacity of the unit cell circuit are improved by using the current input pixel-level ADC. The charge handling capacity of the integrator is appropriately extended to maximize the integration time regardless of the magnitude of the input current and low power supply voltage. The readout circuit was fabricated using a 0.35-μm 2-poly 4-metal CMOS process for a 640 × 512 array with a pixel size of 40 μm × 40 μm. The peak SNR and dynamic range are 77.1 and 80.1 dB, respectively, with a power consumption of 0.62 μW per pixel.

Keywords: pixel-level ADC; current-input ADC; readout circuit; microbolometer; high SNR; wide dynamic range

Citation: Lee, J.; Nam, I.; Woo, D. Current Input Pixel-Level ADC with High SNR and Wide Dynamic Range for a Microbolometer. *Sensors* **2021**, *21*, 2354. https://doi.org/10.3390/s21072354

Academic Editor: Roman Sotner

Received: 17 February 2021
Accepted: 24 March 2021
Published: 28 March 2021

Publisher's Note: MDPI stays neutral with regard to jurisdictional claims in published maps and institutional affiliations.

1. Introduction

Infrared cameras, which detect objects by infrared rays in a situation where they cannot be observed by human eyes, are widely used in military, medical, and commercial fields. In recent years, there has been an increase in the demand for compact and power efficient portable infrared cameras, especially in the private sector for medical devices and security cameras. Therefore, significant research has been conducted on microbolometers, which are uncooled-infrared detectors that do not require a cryogenic cooling system [1–4].

While designing a readout integrated circuit (ROIC) for infrared focal plane array (IRFPA), noise performance is of crucial concern. The dominant noise sources in a microbolometer are the Johnson noise and $1/f$ noise, which can be decreased by increasing the integration time [5]. Therefore, it is necessary to use a pixel-level readout architecture with a large integration capacitor to obtain a long integration time. However, a microbolometer has a high bias current compared to its signal current. Hence, it is hard to locate a large integration capacitor in the pixel because of area limitation. To effectively overcome this problem, a bias-current skimming technique can be used [6]. In addition to noise performance, the dynamic range (DR), which is defined as the ratio of the maximum allowable signal to the minimum detectable signal, is also an important factor. In many applications, either the frame rate is very slow or the target temperature varies over a wide range. Hence, it is necessary to consider both the noise performance and the charge handling capacity of the ROIC.

A monolithic analog-to-digital converter (ADC) is essential for compact and power-efficient portable infrared cameras. Information can be transported with a high signal-to-noise ratio (SNR) in a more power efficient manner in the digital domain using a monolithic ADC, and on-chip signal processing for system-on-chip (SoC) can be made available. Monolithic ADCs can be implemented at the chip level by employing a single high-speed ADC [7], at the column level by using multiple lower speed ADCs [8,9] or at the pixel level by using very low speed ADCs [10–14]. A pixel-level ADC has many advantages

55

over chip and column level ADCs. These include low noise, low power dissipation, and the ability to continuously observe the pixel outputs for long time [14,15].

Although the pixel-level ADC has many advantages, conventional ADC architectures are not easily implemented in IRFPAs because a low-power, high-resolution ADC with small size is required. Since a single-slope ADC (SS-ADC) has a very simple structure, a pixel-level ADC can be easily implemented using this method [10]. However, since analog-to-digital (A/D) conversion is performed after integration, SNR and DR cannot be improved. In addition, a large number of comparisons are required to implement a high-resolution SS-ADC, which significantly increases power consumption.

Another method to implement a pixel-level ADC is based on the time-to-digital (TTD) ADC, also known as the time-to-first spike ADC [11]. While integrating the input current, it detects the time at which the integral signal reaches the reference voltage. Properly modifying the reference voltage can improve the DR characteristics. However, since the integration time varies depending on the input current, it is difficult to obtain a good SNR or a suitable digital resolution for a large input current.

Studies have been conducted that implement pixel-level ADC using pulse frequency modulation (PFM) method [12,13]. While integrating the input current, the integral signal is reset when the integral signal reaches the reference voltage, and A/D conversion is performed using the reset number. Since sufficient integration time and charge handling capacity can be obtained, SNR and DR characteristics can be improved. However, the PFM-based implementation requires many reset pulses to obtain a high-resolution ADC. Therefore, it has high power consumption and is disadvantageous for fast operation. To overcome these problems, the two-stage method can be a good solution [16]. However, since accuracy and linearity are affected by the reset interval, A/D conversion performance may be poor depending on the characteristics of the comparator. In addition, it is still difficult to reduce power consumption since the comparator is always on as the asynchronous method is used.

In the proposed readout circuit, current-input pixel-level ADC using a synchronous two-stage A/D conversion is efficiently performed during the integration, and the characteristics of power consumption, SNR, and DR are improved with high resolution.

2. Basic Concepts of the Proposed ADC

The current input pixel-level ADC proposed in this paper uses a synchronous two-stage A/D conversion to reduce power consumption and improve SNR and DR characteristics. The basic concepts of the proposed ADC are explained in the timing diagram shown in Figure 1. First, the current input is integrated and converted into a voltage $V(C_{INT})$, and it is compared with the reference voltage (V_{TH}) at regular time intervals. V_{RST} and V_{SAT} denote the reset voltage and the maximum allowable voltage of $V(C_{INT})$, respectively, and V_{TH} is equal to $0.5(V_{RST} + V_{SAT})$. If the value of $V(C_{INT})$ is greater than the value of V_{TH} at the comparison point, the constant voltage of $0.5(V_{SAT} - V_{RST})$ is removed from $V(C_{INT})$. This process is repeated periodically during the integration of the current input. The digital information D_{OUT} on the current input I_{SIG} is obtained by counting the number of removals and is expressed by

$$D_{OUT} = \left[\frac{2T_{INT}I_{SIG}}{C_{INT}(V_{SAT} - V_{RST})} \right] \tag{1}$$

where T_{INT} is the integration time, C_{INT} the integration capacitor, and the square bracket [] represents the function of digitization. This method is based on first-order incremental ADC [17]. Using this method, A/D conversion is efficiently performed during integration and a sufficient integration time and large charge-handling capacity is obtained. The maximum charge-handling capacity Q_M is expressed by

$$Q_M = 2^{M-1}C_{INT}(V_{SAT} - V_{RST}) \tag{2}$$

where M is the resolution of the A/D conversion. Since a synchronized method is used where comparing $V(C_{INT})$ and V_{TH}, power consumption of the comparator is minimized.

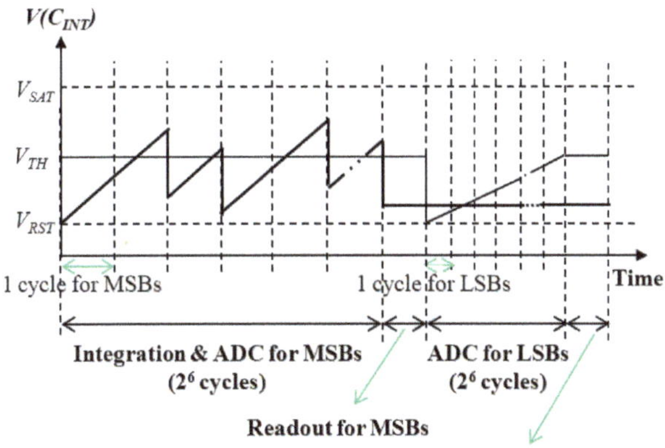

Figure 1. Basic concepts of the proposed pixel-level analog-to-digital converter (ADC).

In a typical application for medium wavelength infrared (MWIR) microbolometer, even if the current input is integrated after removing the bias current, the ADC of the readout circuit must have a resolution of 12-bit or more [12,14]. To implement a 12-bit resolution using only the first-order incremental ADC, 2^{12} comparison cycles are required, which is a great burden in terms of speed and power consumption. To solve these problems, a two-stage A/D conversion is used as shown in Figure 1. In the first period, A/D conversion for the upper six-bit is performed during the integration as described above. After the integration, the residual voltage of $V(C_{INT})$ is held and the lower six-bit is determined by additional single-slope A/D conversion. In this case, since the number of the comparison cycles for implementing a 12-bit ADC is reduced to 2^7 ($2^6 + 2^6$) times, a compact and low-power design is possible, which can be easily implemented as a pixel-level ADC. The circuits for the upper and lower six-bit A/D conversion are shared by dividing the time, so that the conversion result for the upper six-bit should be read out of the array before the lower six-bit conversion.

3. Circuit Implementation

Figures 2 and 3 show the schematic diagram and detailed timing diagram of the proposed pixel-level ADC, respectively. Since a microbolometer has a high bias current compared to its signal current, bias-current suppression is used [6]. Using the accurate bias-current suppression, the integration time can be increased to improve the SNR and the resolution required for A/D conversion can be reduced. Additionally, fixed-pattern noise (FPN), which is produced by the non-uniform IRFPA response, can be lowered to reduce the burden on non-uniformity correction [6,18]. The responsivity of the microbolometer is highly dependent on the bias voltage V_B, hence, maintaining a stable bias voltage is crucial in microbolometer applications [19]. Therefore, a capacitive transimpedance amplifier (CTIA) is used for current integration even though it has a high-power consumption.

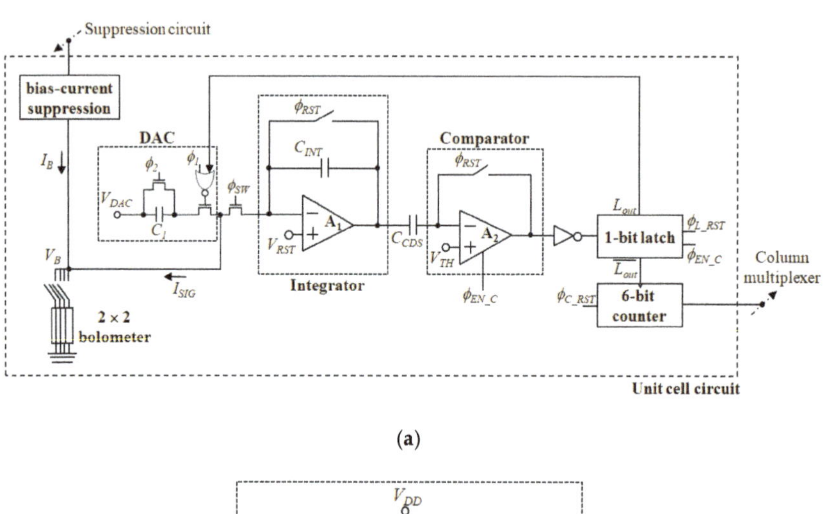

(a)

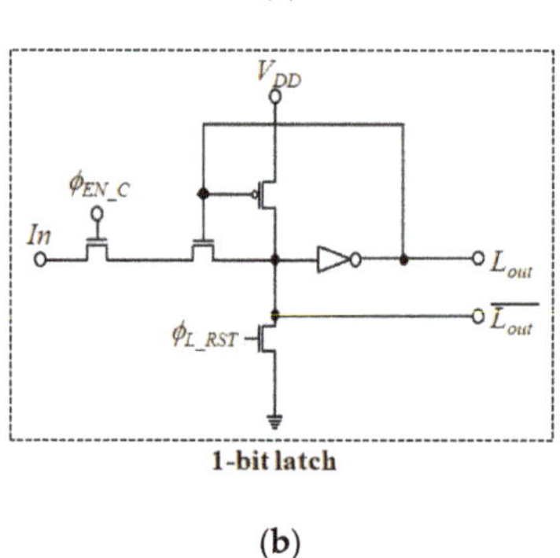

(b)

Figure 2. Schematic diagram of the proposed ADC: (**a**) overall unit cell circuit; (**b**) one-bit latch used in unit cell circuit.

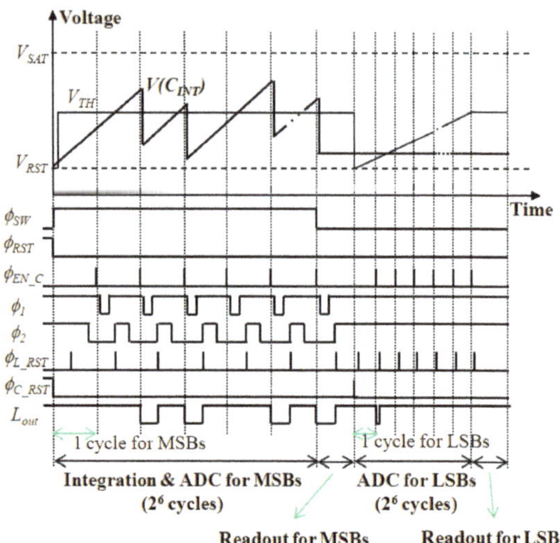

Figure 3. Timing diagram of the proposed ADC shown in Figure 2.

First, the current input I_{SIG} is integrated with the integrator and the value of $V(C_{INT})$ is compared with the value of V_{TH} using the comparator. Here, ϕ_{SW} and ϕ_{RST} are control signals for determining the integration time and reset control signals for the integrator, respectively. The C_{CDS} is the capacitor required for correlated double sampling (CDS) operation [20], reducing FPN and low-frequency noise and compensating for the offset voltage of the comparator and integrator. To completely eliminate all FPNs and non-uniformities in IRFPA and ROIC, an additional calibration process is required. The comparison result is transferred and stored in a one-bit latch according to the ϕ_{EN_C} signal. The result of the comparison is a logic 0', the output of the one-bit latch L_{out} is a logic 0', and the constant charge is removed from the integrator to the DAC according to the control signals of ϕ_1 and ϕ_2. The one-bit latch is then reset according to the ϕ_{L_RST} signal for the next comparison cycle. The inversion signal of L_{out} is used as a clock signal of a six-bit counter, through which the digital value of the upper six bits for the current input is determined. ϕ_{C_RST} is used for the reset control signal of the six-bit counter.

After completing the integration and A/D conversion of the upper six bits, the $V(C_{INT})$ value is held by using ϕ_{SW}, and the digital values stored in the counter are sequentially transferred to the column multiplexer outside the array. A/D conversion of the lower six bits is implemented as a single-slope A/D conversion method for the residual $V(C_{INT})$. First, the six-bit counter is reset using the ϕ_{C_RST}, and the V_{TH} value changes from V_{RST} to $0.5(V_{RST} + V_{SAT})$ in the form of a ramp signal. When the value of V_{TH} is less than the value of $V(C_{INT})$, the input of one-bit latch is logic 1', and the L_{out} changes periodically by ϕ_{EN_C} and ϕ_{L_RST}. Therefore, the digital value of the six-bit counter increases until the V_{TH} exceeds $V(C_{INT})$, and the final digital data stored in the counter is the lower six-bit digital data for the current input. Finally, the six-bit digital data is sequentially transferred once again to the column multiplexer outside the array.

Figure 4 shows a one-bit counter used in the unit cell circuit in Figure 2. It is a simple structure using three logic inverters, and it can be easily applied to the pixel-level ADC. X_{IN} and X_{OUT} are clock and output, respectively, and X_{OUT} turns into the logic value of v_y when X_{IN} changes from logic 1' to 0'. Due to the delay time between X_{IN} and X_{OUT} signals, v_y is not affected by changes in X_{OUT} signal at this moment. Subsequently, when X_{IN} changes from logic 0' to 1', v_y changes to the inversion of X_{OUT}. The final six-bit counter is configured as an asynchronous ripple counter using six one-bit counters in Figure 4.

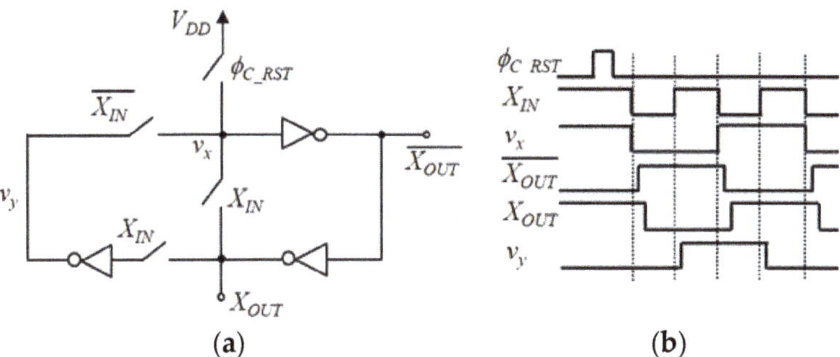

Figure 4. One-bit counter used in the unit cell circuit of Figure 2: (**a**) schematic diagram; (**b**) timing diagram.

Figure 5 shows the overall arrangement of the proposed ROIC. The unit circuit in Figure 2 has a simple configuration. However, it is difficult to implement the proposed unit circuit in one pixel of 40 μm × 40 μm. Thus, as shown in Figure 5, four adjacent (2 × 2 array) bolometers share one unit circuit in Figure 2 in a time-dividing manner. The size of the bolometer array is 640 × 512, and the unit circuit has an array of 320 × 256. The temperature sensor detects the substrate temperature, and the dark sensor obtains the reference value of the bias current. The suppression circuit and 12-bit ADC/DAC blocks are used to control the bias current of the bolometer array [6]. Timing circuit, row scan, and ramp signal generator are required to control the unit circuit, and column multiplexer is required to transfer digital data outside.

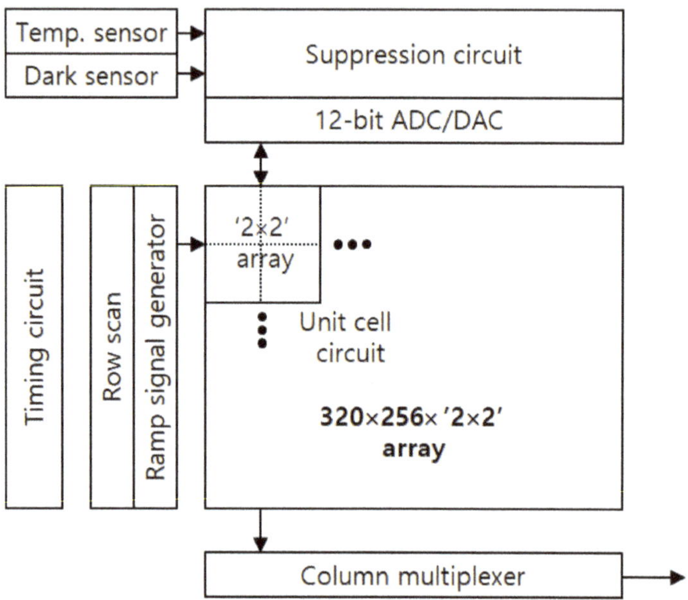

Figure 5. Block diagram of the overall arrangement of the proposed readout integrated circuit (ROIC).

4. Experimental Results

The readout circuit was designed using a 0.35 μm 2-poly 4-metal CMOS process for a 640 × 512 MWIR a-Si microbolometer array with a pixel size of 40 μm × 40 μm. The design parameters of the microbolometer and readout circuit are summarized in Table 1. The fabricated unit cell circuit is 80×80 μm^2 in size because the proposed unit circuit is shared by 2 × 2 pixels. Uncooled microbolometers are difficult to operate at speeds above 60Hz. Therefore, the frame rate of IRFPA is set to 60 Hz and the sampling rate of the proposed pixel-level ADC is 240 Hz. It was tested at room temperature using the operation board shown in Figure 6.

Table 1. Design parameters of the microbolometer and readout circuit.

Parameter	Value
Resistance	10 MΩ
Temperature coefficient of resistance	2.5%
Thermal conductance	5×10^{-8} W/K
Fill factor	80%
Emissivity	80%
Frame rate	60 Hz
Average bias current (I_B)	40 nA
Maximum signal current (I_{SIG})	20.4 nA
Integration time (T_{INT})	3.1 ms
Integration capacitor (C_{INT})	1 pF
Power consumption	<0.62 μW

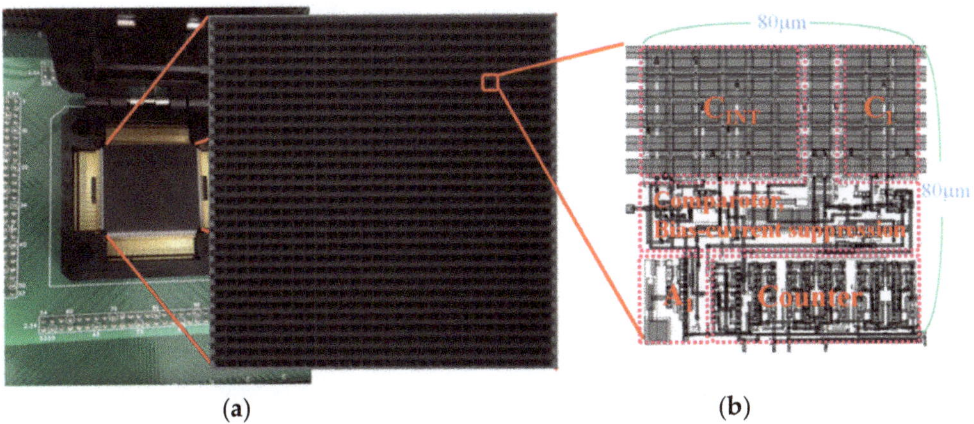

(a) **(b)**

Figure 6. Measurement system: (**a**) operation board and fabricated chip; (**b**) mask layout of the proposed unit cell circuit for 2×2 bolometer array.

Figure 7 shows the operating waveforms of the proposed pixel-level ADC. The current input is integrated and converted to MSBs in the first period. Subsequently, the residual $V(C_{INT})$ is held and converted to LSBs. MSBs and LSBs are estimated by counting the number of $\overline{L_{OUT}}$. Figure 7b shows the operating waveforms of the counter output ($Q_3 Q_2 Q_1 Q_0$). Only four bits are displayed among six-bit outputs, and Q_0 represents the least significant bit. As shown in Figure 7, the range of the available current input that does not saturate during integration time is drastically improved by the current input pixel-level ADC.

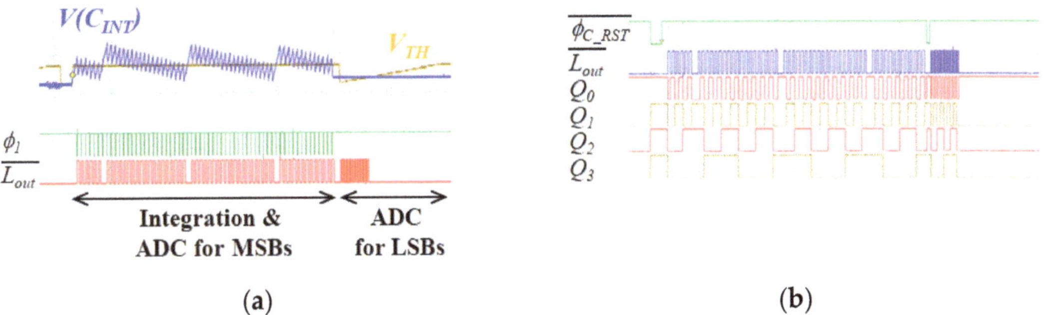

(a) **(b)**

Figure 7. Measured operating waveforms of the proposed pixel-level ADC: (**a**) current integration and A/D conversion; (**b**) counter output.

Figure 8a shows the measurement results of differential non-linearity (DNL). Figure 8b shows the integral non-linearity (INL) estimated by Figure 8a. The DNL is within about ± 0.4 LSB, and the INL is within about ± 2.0 LSB. These results confirm the monotonicity of the proposed pixel-level ADC.

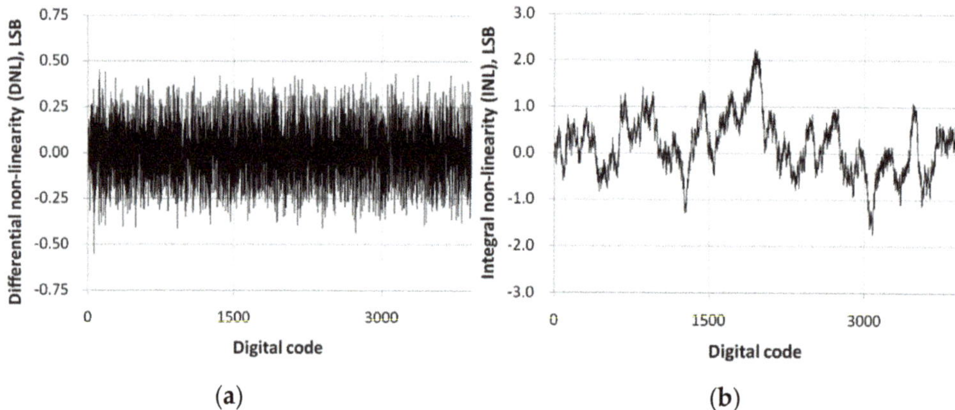

Figure 8. Measurement results of the linearity of the proposed pixel-level ADC: (**a**) differential non-linearity (DNL); (**b**) integral non-linearity (INL).

SNR and DR characteristics are important figures of merit for IRFPA applications. To estimate the SNR and DR characteristics, the readout circuit noise, i.e., the ADC noise, should be measured. First, the digital output of the ADC for a zero current input was sampled 2000 times, then the readout noise for a zero current input was estimated as the standard deviation σ of the sampled output. Through this process, it was confirmed that the readout noise for a zero-current input σ_0 is close to 0 LSB. However, this method alone cannot evaluate the effect of kTC noise of the DAC on the readout noise. In order to evaluate the readout noise caused by the DAC, there must be a current input. In this case, the readout noise itself can be evaluated by eliminating the noise caused by the current input. The ADC output was sampled several times for two different current inputs, and the histograms were obtained as shown in Figure 9. Figure 9a,b are histograms obtained using a total of 2000 samples, where the number of MSBs counting is 1 and 63, respectively. Using the data in Figure 9, the noise of the current input system σ_I can be removed, and the noise equivalent to one MSB counting σ_M can be predicted as shown in following equation:

$$\sigma_a^2 = \sigma_0^2 + \sigma_I^2 + \sigma_M^2 \tag{3}$$

$$\sigma_b^2 = \sigma_0^2 + \sigma_I^2 + 63\sigma_M^2 \tag{4}$$

$$\sigma_M = \sqrt{(\sigma_b^2 - \sigma_a^2)/62} = 0.0438 \,[\text{LSB}] \tag{5}$$

where σ_a and σ_b are the σ of Figure 9a,b, respectively. After predicting the σ_M value from the histograms of Figure 9, the readout noise according to the current input can be evaluated and applied to the final SNR characteristic of Figure 10.

Figure 10 shows the SNR characteristics according to the current input, and the total noise is estimated from the bolometer noise, quantization noise, and the readout noise. The legend A of Figure 10 indicates the graph of the proposed pixel-level ADC with a T_{INT} of 3.1 ms and a maximum I_{SIG} of 20.4 nA. Legends B and C indicate the case of applying the conventional 12-bit SS-ADC after the integration time. In the case of B and C, since the charge handling capacity is limited due to the pixel area limitation, it is necessary to reduce the integration time to 98 μs for a wide current input (B) or reduce the input range to 3.0 nA for a long integration time (C). Design parameters except for T_{INT} and maximum I_{SIG} are the same as in Table 1 for A, B, and C. Since the charge handling capacity of the proposed ADC is appropriately extended, the integration time is maximized for a wide input range. Therefore, the proposed ADC has good SNR characteristics for a wide input range as shown in Figure 10.

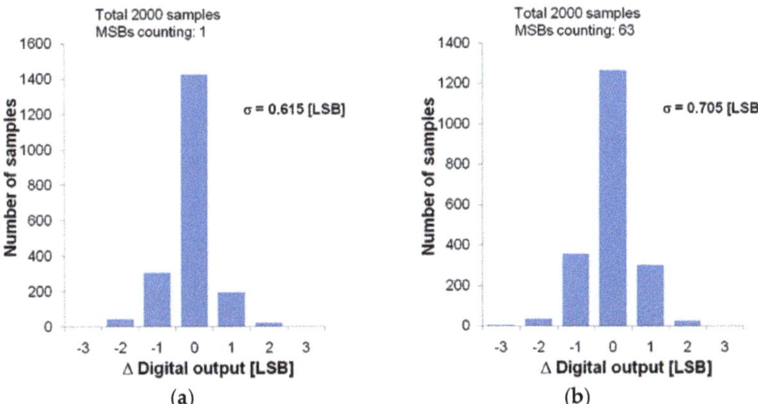

Figure 9. Histograms of the digital output of the proposed ADC with a fixed input: (**a**) where the number of MSBs counting is 1; (**b**) where the number of MSBs counting is 63.

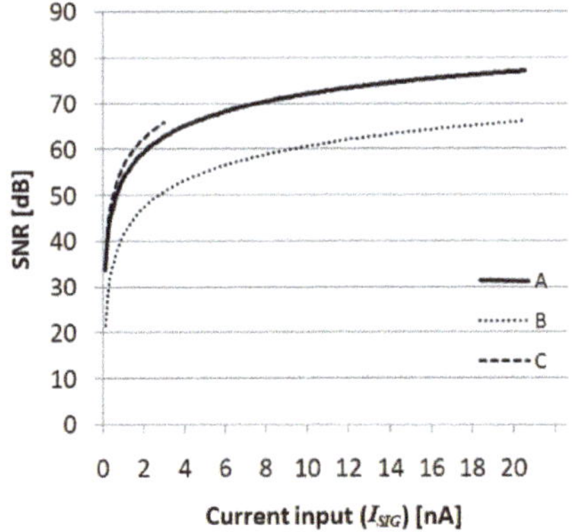

Figure 10. SNR characteristics for the three cases: (**A**) the proposed pixel-level ADC; (**B**) 12-bit SS-ADC with reduced integration time; and (**C**) 12-bit SS-ADC with reduced input range.

Table 2 shows the performance comparison between the proposed ADC and other methods. The design of the readout circuit for IRFPA is greatly dependent on the IR detector. Therefore, in the case of A, A′, B, and C, the main design parameters are the same for reliable comparison. A′ indicates the case of applying the proposed ADC without bias-current suppression. Since the maximum value of the bias current is about 60 nA, if the bias current is not suppressed, the integration time decreases and both charge handling capacity and SNR decrease. Without bias-current suppression, the effective signal range decreases and the burden required for non-uniformity correction is increased [12,14]. Therefore, the resolution of the ADC must be increased in the actual design. Although it is difficult to make a clear comparison, the results presented in other papers are compared Table 2. Reference [10] uses the SS-ADC for pixel-level ADC and shows very low charge-handling capacity like B and C. Since the readout circuit of the reference [10] is for a cooled type MWIR detector, the bias current is small and the readout noise is very low. In addition, since the SNR value of A includes the noise of the detector, it is difficult to clearly compare the SNR values of A and [10]. The reference [11] uses the TTD method for pixel-level ADC, and the proposed ADC shows superior SNR characteristics. Since the readout circuit of the reference [11] is for a short wavelength infrared (SWIR) detector, an additional method is used to increase the charge handling capacity to an extreme. It can be seen from the

Table 2 that the proposed ADC shows excellent and balanced characteristics in both charge handling capacity and SNR.

Table 2. Performance comparison.

	Charge Handling Capacity [e-]	Peak SNR [dB]	Description
A	405 M	77.1	Proposed ADC
A'	101M	74.6	Proposed ADC without Bias-current suppression
B	12.5 M	66.2	SS-ADC with reduced T_{INT}
C	59.6 M	66.0	SS-ADC with reduced I_{SIG} range
[10]	10 M	89.6	SS-ADC
[11]	1.98 G	60.9	TTD
[12]	310 M	>60	PFM

Figure 11 is a comparison of the results of the power consumption for the three cases, simulated using the same MOS library. The three cases are designed based on the same microbolometer shown in Table 1, and the bias-current suppression method is applied to all three circuits. Legend A indicates the power consumption of the proposed ADC, and D indicates the case of applying a PFM-based method to the A/D conversion for the upper six bits. Legend E indicates a case where the entire A/D conversion is implemented with only first-order incremental ADC. Considering the characteristics of the comparator and the maximum integration time, only 10-bit conversion is applied instead of 12 bits in the case of legend E. Since the comparator is always on as the asynchronous method is used in the case of D, it is still difficult to reduce power consumption. The 1st-order incremental ADC requires many comparison cycles to obtain a high-resolution ADC. Therefore, it has high power consumption. The proposed ADC can expect the lowest power consumption among the three methods as shown in Figure 11.

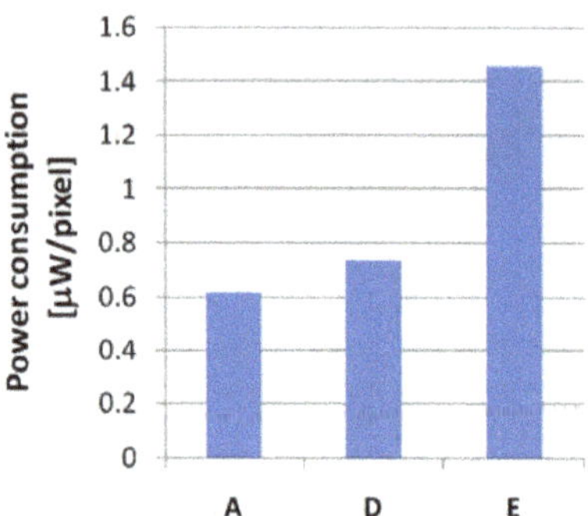

Figure 11. Power consumption for the three cases: (**A**) the proposed pixel-level ADC; (**D**) pulse frequency modulation (PFM)-based two-stage analog-to-digital (A/D) conversion; (**E**) 1st-order incremental ADC.

5. Conclusions

The current-input pixel-level ADC with a high SNR and wide DR is proposed for the readout circuit of the MWIR microbolometer array. In the proposed readout circuit, A/D conversion is efficiently performed during the integration, and a sufficient integration time and large charge-handling capacity is obtained. Using a synchronous two-stage A/D conversion, the characteristics of the power consumption, SNR, and DR are improved with high resolution. The proposed pixel-level ADC has a very simple structure and low power consumption. This enables the final readout circuit to be suitable for applications in portable infrared cameras.

Author Contributions: D.W. supervised the research. J.L. and D.W. proposed the idea and designed the circuit. J.L. performed the simulation and mask layout. J.L. and I.N. performed the experiment and analysis. J.L. and D.W. wrote the initial manuscript. D.W. and I.N. revised and finalized the paper. All authors have read and agreed to the published version of the manuscript.

Funding: This research was funded by the Korea government (MSIT), grant number 2020R1F1A1052571.

Institutional Review Board Statement: Not applicable.

Informed Consent Statement: Not applicable.

Data Availability Statement: Not applicable.

Acknowledgments: This work was supported by the National Research Foundation of Korea (NRF) grant funded by the Korea government (MSIT) (No. 2020R1F1A1052571). This work was supported by the Catholic University of Korea, Research Fund, 2019. The chip fabrication and EDA Tool were supported by the IC Design Education Center.

Conflicts of Interest: The authors declare no conflict of interest.

References

1. Akin, T. Low-Cost LWIR-Band CMOS Infrared (CIR) Microbolometers for High Volume Applications. In Proceedings of the IEEE 33rd International Conference on MEMS, Vancouver, BC, Canada, 18–22 January 2020; pp. 147–152.
2. Abbasi, S.; Shafique, A.; Ceylan, O.; Gurbuz, Y. A Digital Readout IC for Microbolometer Imagers Offering Low Power and Improved Self-Heating Compensation. *IEEE Sens. J.* **2020**, *20*, 909–917. [CrossRef]
3. Yen, P.; Su, G.J. Uncooled Long-Wavelength Infrared Sensors Using Cytochrome C Protein on Suspension Electrodes with CMOS Readout Circuits. *IEEE Sens. J.* **2019**, *19*, 10221–10227. [CrossRef]
4. Jiang, Z.; Men, L.; Wan, C.; Xiao, P.; Jiang, C.; Tu, X.; Jia, X.; Kang, L.; Li, L.; Chen, J.; et al. Low-Noise Readout Integrated Circuit for Terahertz Array Detector. *IEEE Trans. Terahertz Sci. Technol.* **2018**, *8*, 350–356. [CrossRef]
5. Jakonis, D.; Svensson, C.; Jansson, C. Readout architectures for uncooled IR detector arrays. *Sens. Actuators* **2000**, *84*, 220–229. [CrossRef]
6. Woo, D.H.; Kang, S.G.; Lee, H.C. Novel current mode background suppression for 2-D LWIR application. *IEEE Trans. Circuits Syst.* **2005**, *52*, 606–610.
7. Kang, S.G.; Woo, D.H.; Lee, H.C. Multiple integration method for a high signal-to-noise ratio readout integrated circuit. *IEEE Trans. Circuits Syst. Express Briefs* **2005**, *52*, 553–557. [CrossRef]
8. Wang, G.; Lu, W.; Zhang, L.; Zhang, Y.; Chen, Z. 14-bit 20 μW column-level two-step ADC for 640 × 512 IRFPA. *Electron. Lett.* **2015**, *51*, 1054–1056. [CrossRef]
9. Liu, D.; Lu, W.; Chen, Z.; Zhang, Y.; Lei, S.; Tan, G. A 14-bit differential-ramp single-slope column-level ADC for 640 × 512 uncooled infrared imager. *IEEE Int. Symp. Circuits Syst.* **2016**, *22*, 1922–1925.
10. Huang, Z.; Niu, Y.; Lu, W.; Chen, G.; Li, Y.; Zhang, S.; Chen, Z. A 16-bit Single-Slope based Pixel-level ADC for 15μm-pitch 640×512 MWIR FPAs. *IEEE Int. Symp. Circuits Syst.* **2018**, *27*, 1–5.
11. Jo, Y.M.; Woo, D.H.; Kang, S.G.; Lee, H.C. Very Wide Dynamic Range ROIC With Pixel-Level ADC for SWIR FPAs. *IEEE Sens. J.* **2016**, *16*, 7227–7233. [CrossRef]
12. Tchagaspanian, M.; Villard, P.; Dupont, B.; Chammings, G.; Martin, J.L.; Pistre, C.; Lattard, D.; Chantre, C.; Arnaud, A.; Yon, J.J.; et al. Design of ADC in 25 μm pixels pitch dedicated for IRFPA image processing at LETI. In *Infrared Technology and Applications*; International Society for Optics and Photonics: Bellingham, WA, USA, 2007; Volume 6542, pp. W1–W65421.
13. Li, M.; Li, F.; Zhang, C.; Wang, Z. Pixel-level A/D conversion using voltage reset technique. *J. Semicond.* **2014**, *35*, 115009–1–5. [CrossRef]
14. Hwang, C.H.; Woo, D.H.; Lee, H.C. Pixel-level ADC with two-step integration for 2-D microbolometer IRFPA. *Ieice Trans. Electron.* **2011**, *94*, 1909–1912. [CrossRef]
15. Pain, B.; Fossum, E.R. Approaches and analysis for on-focal-plane analog-to-digital conversion. In *Infrared Readout Electronics II*; International Society for Optics and Photonics: Bellingham, WA, USA, 1994; Volume 2226, pp. 208–218.
16. Yang, S.; Luo, L.; Lu, W.; Shi, G.; Lyu, B.; Niu, Y.; Zhang, Y.; Chen, Z. A 16-bit two-step pixel-level ADC for 384*288 infrared focal plane array. In Proceedings of the IEEE 12th International Conference on ASIC (ASICON), Guiyang, China, 28 October 2017; pp. 780–783.
17. Robert, J.; Valencic, V. Offset and Charge Injection Compensation in an Incremental Analog to Digital Converter. In Proceedings of the 11th European Solid-State Circuits Conference (ESSCIRC), Montreux, Switzerland, 16–18 September 1985; pp. 45–48.
18. Jo, Y.M.; Woo, D.H.; Lee, H.C. TEC-Less ROIC With Self-Bias Equalization for Microbolometer FPA. *IEEE Sens. J.* **2015**, *15*, 82–88.
19. Hwang, C.H.; Kim, C.B.; Lee, Y.S.; Lee, H.C. Pixelwise readout circuit with current mirroring injection for microbolometer FPAs. *Electron. Lett.* **2008**, *44*, 732–733. [CrossRef]
20. Nitta, Y.; Muramatsu, Y.; Amano, K.; Toyama, T.; Yamamoto, J.; Mishina, M.; Suzuki, A.; Taura, T.; Kato, A.; Kikuchi, M.; et al. High-Speed Digital Double Sampling with Analog CDS on Column Parallel ADC Architecture for Low-Noise Active Pixel Sensor. In Proceedings of the IEEE International Solid State Circuits Conference, Francisco, CA, USA, 5–9 February 2006; pp. 2024–2031.

Article

A Dynamic Threshold Cancellation Technique for a High-Power Conversion Efficiency CMOS Rectifier

António Godinho [1,†], Zhaochu Yang [1,†], Tao Dong [2,*], Luís Gonçalves [3], Paulo Mendes [3], Yumei Wen [4], Ping Li [4] and Zhuangde Jiang [1]

1   Chongqing Key Laboratory of Micro-Nano Systems and Smart Transduction, Collaborative Innovation Center on Micro-Nano Transduction and Intelligent Eco-Internet of Things, Chongqing Key Laboratory of Colleges and Universities on Micro-Nano Systems Technology and Smart Transducing, National Research Base of Intelligent Manufacturing Service, Chongqing Technology and Business University, Chongqing 400067, China; godinhoa2208@gmail.com (A.G.); zhaochu.yang@ctbu.edu.cn (Z.Y.); zdjiang@mail.xjtu.edu.cn (Z.J.)
2   Department of Microsystems (IMS), Faculty of Technology Natural Sciences and Maritime Sciences, University of South-Eastern Norway, 3616 Kongsberg, Norway
3   Center for MicroElectromechanical Systems (CMEMS-UMinho), University of Minho, 4800-058 Guimarães, Portugal; lgoncalves@dei.uminho.pt (L.G.); paulo.mendes@dei.uminho.pt (P.M.)
4   School of Electronic Information and Electrical Engineering, Shanghai Jiao Tong University, Shanghai 200240, China; yumei.wen@sjtu.edu.cn (Y.W.); liping_sh@sjtu.edu.cn (P.L.)
*   Correspondence: tao.dong@usn.no
†   Authors whom first authorship is shared by António Godinho and Zhaochu Yang.

Citation: Godinho, A.; Yang, Z.; Dong, T.; Gonçalves, L.; Mendes, P.; Wen, Y.; Li, P.; Jiang, Z. A Dynamic Threshold Cancellation Technique for a High-Power Conversion Efficiency CMOS Rectifier. *Sensors* **2021**, *21*, 6883. https://doi.org/10.3390/s21206883

Academic Editors: Jong-Ryul Yang and Seong-Tae Han

Received: 24 September 2021
Accepted: 13 October 2021
Published: 17 October 2021

Publisher's Note: MDPI stays neutral with regard to jurisdictional claims in published maps and institutional affiliations.

Abstract: Power conversion efficiency (PCE) has been one of the key concerns for power management circuits (PMC) due to the low output power of the vibrational energy harvesters. This work reports a dynamic threshold cancellation technique for a high-power conversion efficiency CMOS rectifier. The proposed rectifier consists of two stages, one passive stage with a negative voltage converter, and another stage with an active diode controlled by a threshold cancellation circuit. The former stage conducts the signal full-wave rectification with a voltage drop of 1 mV, whereas the latter reduces the reverse leakage current, consequently enhancing the output power delivered to the ohmic load. As a result, the rectifier can achieve a voltage and power conversion efficiency of over 99% and 90%, respectively, for an input voltage of 0.45 V and for low ohmic loads. The proposed circuit is designed in a standard 130 nm CMOS process and works for an operating frequency range from 800 Hz to 51.2 kHz, which is promising for practical applications.

Keywords: vibration energy harvester; power management circuit; CMOS rectifier; dynamic threshold cancellation technique; high power conversion efficiency

1. Introduction

Presently, energy harvesting appears as a promising reliable technology that can prolong the lifetime of batteries and power wireless sensor networks (WSNs) for environmental monitoring [1]. However, in these WSN applications, ambient vibrations are unpredictable, time-varying, and low amplitude, which restricts the available power of the energy harvesting system [2]. To overcome these drawbacks, research groups have been focusing on using piezoelectric harvesters due to their high power density and capability to integrate MEMS and CMOS technology, making it possible to develop all the systems (energy harvester and electronic system) in a single chip [3–5]. Thus, to maximize the amount of energy transferred under different ambient conditions, a power management circuit (PMC) is crucial in order to extract, convert, store, regulate, and manage the scavenged energy from the piezoelectric device [6,7].

Because the vibrational energy sources produce AC signals, scavenging such energy requires a full-wave rectifier as a key circuit inside the PMC, which allows the AC/DC

67

conversion to properly power the WSNs. However, because the output power of the vibrational energy harvester is low [4], the high forward voltage required by standard full-wave diode bridges and Schottky diode rectifiers limits their use on these low power restrict applications [8]. To surpass these limitations, diode-connected MOS transistors have been widely used because they present similar I-V characteristics to the standard diodes. Thus, designing the rectifier in CMOS technology is highly desirable to decrease the device's form factor and easily integrate with the energy harvester while exploring new dynamic techniques to reduce the power consumption, achieve high PCE, and minimize leakage current [9,10].

Recent work has been developing dynamic threshold techniques to reduce the threshold voltage effect [11]. Addressing these techniques allows for the reduction of total voltage drop and mitigation of the reverse leakage current in the active stage. By attending to these concerns during the design of the circuit, it is possible to minimize the circuit's overall power and leakage current consumption. Thus, all these conditions were carefully considered during the design of the proposed high-power efficiency CMOS rectifier to attend to the demands of this application.

In this work, a new CMOS rectifier structure for piezoelectric energy harvesters is presented. It combines a passive stage negative voltage converter (NVC) with an active diode controlled by a dynamic threshold cancellation circuit to build a new architecture that can reduce its total voltage drop. With this configuration, a voltage drop lower than 2 mV can be achieved in the second stage, which consequently enhances features such as VCE and PCE, as well as reduces the reverse leakage current that flows from the load.

2. CMOS Rectifiers

2.1. Passive Rectifiers

The CMOS gate cross coupled can replace the conventional full-wave bridge rectifier to overcome the high forward voltage drop because it allows a minimum input voltage to operate [10,12,13]. However, this topology still lacks efficiency due to the threshold voltage (V_{TH}) drop across the diode connected in each conduction path [13].

The fully cross-coupled rectifier intends to fulfill the gap of the previous configuration by eliminating all V_{TH} drops, which reduces the voltage drop across this stage [13]. Consequently, this topology improves both PCE and VCE of the circuit [13]. However, the reverse leakage current appears to be the main disadvantage of using this single configuration, which affects the power transferred from the circuit to the load [10]. Thus, an extra circuit must be added to overcome this issue.

2.2. Active Rectifiers

To prevent the circuit from reverse leakage current, the CMOS passive rectifiers combined with an active configuration can mitigate the reverse leakage current to enhance the DC power of the load [14–19]. In these active configurations, comparators are designed to control the gate voltage of the active diode (or so-called the main transistor) depending on its input and output voltage conditions. In work done by Peters et al. [15], an active rectifier with a bulk-input comparator technique is proposed for ultra-low-voltage energy harvesting systems. However, when the input voltage is higher than the output voltage, the PN junctions between the bulk and source terminal of the input transistor will be turned on. Consequently, the reverse leakage current will flow from the cathode terminal to the anode terminal through the body PN junctions, which compromises the efficiency of the circuit [8]. In addition, the proposed rectifier in [19] has a frequency range not suitable for the application of this research work. In contrast, in the following research papers [8,9,20,21], the frequency bandwidth corresponds to the desired application. The authors use two active diodes to control the reverse current that flows through the two NMOS in each input cycle, and two PMOS in cross-coupled to provide the conduction path. However, the dynamic range does not meet the requirements to achieve a high PCE for input voltages lower than 1 V, which is critical for energy harvesting applications [8].

In [20], the authors designed a fully active configuration using PMOS and NMOS to ensure that the reverse current through the PMOS input source is zero. The main disadvantage of this configuration occurs when the two NMOS devices turn on simultaneously, which leads to power losses. Chang et al. [21] proposed a rectifier with a third comparator to eliminate the oscillations of NMOS, which avoids the two active diodes turning on/off simultaneously. Nevertheless, the PCE is only high for an input voltage around 4.88 V.

However, the main limitations of these configurations are that they cannot control the V_G of the main transistor to increase V_{SG} during the conduction phase. Thus, it is not possible to reduce the internal resistance of this transistor, which limits the output power of the rectifier. Therefore, an extra circuit is needed to reduce the threshold voltage effect of this transistor to overcome these drawbacks.

2.3. Threshold Cancellation Topologies

Several threshold cancellation topologies were proposed to enhance the output stored voltage by dynamically reducing the threshold voltage effect of the main transistor of the rectifier [22–25]. The threshold voltage is a process parameter dependent on the oxide type and thickness [24]. Low threshold voltage MOSFETs present a high leakage current caused by the low substrate doping, which leads to an increase in power consumption and reliability problems [24,26]. Thus, these threshold cancellation techniques are used to avoid those types of MOSFETs since it is only needed to reduce the threshold voltage effect when the main pass transistor is ON. In [25], a low-voltage CMOS rectifier is proposed to perform this technique by using the bootstrap technique, which has enhanced the output voltage stored in the load capacitor. However, for the minimum operating voltage of this configuration (0.8 V), the PCE of this circuit is around 30%, which is not enough for the requirements of this application.

An active bootstrapping rectifier is presented in [27] to overcome the issues of the previous work. This topology uses two active diodes to control the conduction path for each input cycle and a bootstrap technique to reduce the threshold voltage of both main pass PMOS. Additionally, an adaptive voltage converter is set in this work to adjust the gate voltage of the main pass PMOS, which reduces the voltage drop by reducing the on-resistance. Besides lowering the reverse leakage current, the PCE of this configuration can still be improved for input voltages smaller than 1 V. To overcome the low PCE values for a narrow input voltage range, in [28], a dual switching technique replaced the two active diodes. This approach can maintain a constant gate bias on the two main NMOS transistors, avoid the reverse leakage current, reduce the area on-chip, and enhance the PCE for low voltage applications. However, high values for PCE can only be obtained for input frequencies around 20 kHz, which makes the frequency bandwidth narrow.

3. Design Implementation

Regarding the inherent output characteristics of the piezoelectric transducer, the proposed CMOS rectifier was mainly designed to achieve a high PCE for wide low input voltage and frequency conditions. Therefore, the operational voltage ranges from 0.4 V to 1 V, and the working frequency varies from hundreds of Hz to a few kHz. In addition, the output impedance of the energy harvester is not considered in this design because the matching impedance process is performed before this rectification stage in the PMC. Thus, the main goal of this work is to reduce the voltage drop across the structure by applying a threshold cancellation technique that will further enhance the power converted to the ohmic load. These improvements will overcome the drawbacks of previous work by mitigating the reverse leakage current, and thus enhancing the PCE for a low input voltage range.

Figure 1 shows the simplified schematic of the proposed active rectifier. It consists of an NVC and an active diode biased by a threshold cancellation circuit. The first stage is set to perform the signal full-wave rectification. However, because this passive stage cannot control the reverse current from the load capacitor when the output voltage is higher than the input, a second stage active diode (M5) is needed. This active stage is composed of a PMOS controlled by a threshold cancellation circuit with a bootstrapping capacitor to reduce the effective threshold of the active diode, and an adaptive voltage controller (AVC) to adjust the gate voltage of M5 by controlling the charging/discharging cycle of the bootstrapping capacitor. To perform it, a two-input common gate comparator and an NMOS transistor are used. Besides these stages, a dynamic switching bulk (DSB) technique was used to control the bulk voltage of the active diode PMOS.

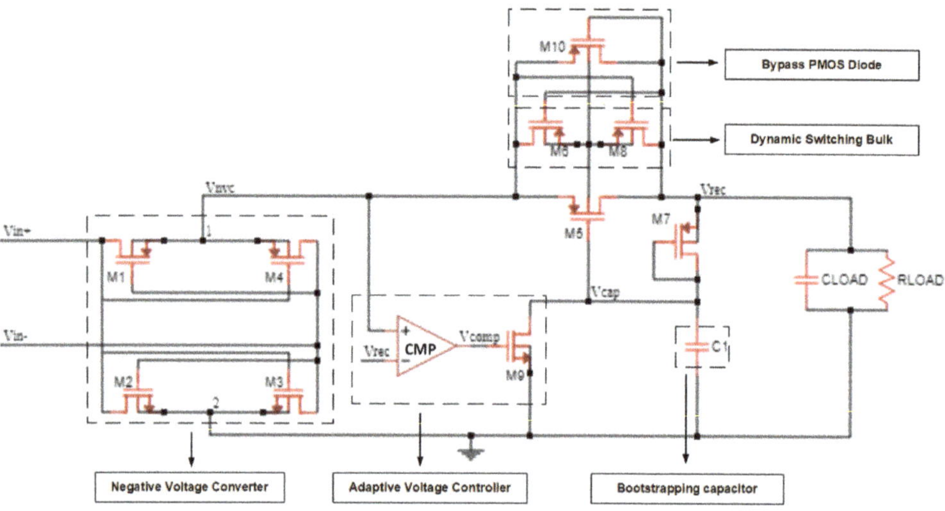

Figure 1. Schematic of the proposed active rectifier composed by a NVC and an active diode controlled by a threshold cancellation circuit.

3.1. Negative Voltage Converter

The first stage is fully passive, and it is used to perform the signal full-wave rectification by applying a fully-cross coupled configuration. During the positive half period of the input signal ($V_{in+} > V_{in-}$), M1 and M3 will be conductive as soon as the input voltage gets larger than V_{THn} and $|V_{THp}|$. In this cycle, node 1 is connected to V_{in+} and node 2 to V_{in-}. For the negative period of the sine wave, M2 and M4 are conducting while the previous two transistors are now turned off (cut-off region). Therefore, the higher voltage potential is always at V_{nvc}, whereas the lowest potential is at 0 V. The voltage drop of the NVC is given by $V_{DSn} + V_{SDp}$ in each conduction path, where V_{DSn} and V_{SDp} are the voltage drop of NMOS transistors M2 or M3 and PMOS transistors M1 or M4, respectively.

To meet all the power restrictions related to the piezoelectric energy harvesting systems, the rectifier circuit must minimize the voltage drop across the rectification process. As less voltage drop occurs, both the VCE and the PCE of the circuit will be higher. For this stage, NVC, the main requirement is to decrease the voltage drop associated with each MOSFET by reducing their on-resistance.

3.2. Active Diode

One of the main challenges on the rectifier circuit is to avoid the reverse leakage current by controlling the operation of transistor M5. Therefore, an active diode controlled with a threshold cancellation circuit can regulate the work behavior of this device depending on the voltage potential between the input and output. The deployed threshold cancellation circuit controls the gate potential of the MOSFET M5 by comparing the input/output voltage conditions. Additionally, the width of M5 has a large influence on the performance of this rectifier because the voltage drop is mainly affected by this parameter due to the internal on-resistance. Consequently, since the gate capacitance of M5 depends on the width, the turn on/off time of the transistor will also be affected by this parameter. In addition, the DSB technique, composed of M6 and M8, is deployed to reduce the leakage current through the bulk terminal of M5 by connecting it to the higher potential (V_{nvc} or V_{rec}). Another advantage of this technique is eliminating the body effect of M5, which reduces the rectifier voltage drop. Both M6 and M8 can be small in size since only a very low current flows through them during the start-up phase.

To assure a safe start-up of M5, a bypass PMOS diode (M10) was connected in parallel. This transistor makes the active diode more robust by preventing it from leakage current in the subtraction that induces latch-up. After the start-up phase, the bypass diode always operates in the cut-off region.

3.3. Threshold Cancellation Circuit

In order to reduce the threshold voltage effect on M5, a bootstrap technique is used by attaching the capacitor C_1 to the output terminal. When the V_{NVC} is higher than the output voltage V_{rec}, M5 is turned ON, since V_{SG5} is no longer lower than V_{TH5}, and thus it can be defined in (1). Nevertheless, because M5 is operating in the deep-triode region due to $V_{SD5} \ll 2 \cdot (V_{SG5} - |V_{TH5}|)$, V_{SG5} can also be defined according to the on-resistance equation, see (2).

$$V_{SG5} = V_{NVC} - V_{CAP} \tag{1}$$

$$V_{SG5} = \frac{1}{\mu_p \cdot C_{ox} \cdot W_5/L_5 \cdot R_{SD5}} + |V_{TH5}| \tag{2}$$

Here, μ_p is the carrier mobility, C_{ox} is the oxide capacitance, W_5/L_5 is the aspect ratio of transistor M5, and V_{TH5} is its respective threshold voltage.

The bootstrapping capacitor (C_1) is charged up through an auxiliary diode-connected PMOS transistor M7, and it maintains a value when the rectifier is under the steady-state regime. At this time, because C_1 is discharging, V_{CAP} is one diode forward-bias voltage (V_{TH7}) bellow V_{rec} due to M7 is being in the saturation region. Thus, the voltage held on the bootstrapping capacitor can be defined as:

$$V_{CAP} = V_{rec} - |V_{TH7}| \tag{3}$$

V_{SG5} and V_{CAP} from (2) and (3), respectively, can be replaced in (1), which means that V_{rec} can now be defined according to the following equation:

$$V_{rec} = V_{NVC} - (|V_{TH5}| - |V_{TH7}|) - \frac{1}{\mu_p \cdot C_{ox} \cdot W_5/L_5 \cdot R_{SD5}} \tag{4}$$

According to (4), the rectified signal is highly influenced by the size of M5 and the threshold voltage of both M5 and M7, and thus it is vital to manage these parameters to enhance the output signal voltage. The implemented threshold cancellation circuit reduces the voltage drop of the main pass transistor M5 by lowering the threshold voltage effect. Additionally, the size of the bootstrap capacitor is an important design concern for the implementation of the proposed rectifier. Integrated capacitors consume a large area on the chip when standard CMOS processes are used [24]. Therefore, C_1 was set at 200 fF not only to reduce the correspondent area on the die but also to have a faster charging/discharging time. Consequently, this low bootstrap capacitance allows a lower gate voltage of M5 at the ON state. Due to the reduction of its internal source to drain resistance, the voltage drop is decreased. The reverse leakage current during the OFF state will be avoided because V_{SG5} is reduced. Moreover, it is necessary to have an auxiliary circuit to hold the V_{CAP} node when M5 is OFF, and to discharge it at the opposite state.

The bootstrapping capacitor is used to reduce the threshold voltage effect of M5. However, an increase in its on-resistance can be noticed due to the reduction of V_{SG5}. Thus, a conduction path needs to be generated to discharge the gate of M5 during the ON state, which will lead to a further increase of V_{SG5}. The proposed AVC is composed of NMOS M9 and a comparator CMP that drives its gate. When V_{NVC} is higher than the output voltage V_{rec}, the comparator CMP should immediately turn on M9 to provide a discharge path of the V_{CAP} node. Consequently, it will turn on the main pass transistor M5 with a low on-resistance. Because the large size of M5 increases the gate capacitance, the AVC must have a faster bias signal control to switch the discharge path of the gate node (V_{CAP}). Thus, the comparator must be designed to attend to these demands.

Figure 2 shows the proposed two-input common gate comparator. This comparator is composed of a current mirror stage to make the comparison, plus an inverter block to bias the gate of M9. Even if the transistor of the current mirror should be as small as possible to reduce the current consumption of the comparator, the size of M12 and M15 must be carefully chosen to manage the delay, and consequently, the reverse leakage current in M5. These two transistors cannot have the same W/L ratio as M11 and M14. Otherwise, this would generate a delay caused by the inverter's gate capacitance's low charging/discharging time. Additionally, they cannot be much larger than the other transistors because of the reduced time that M5 would be ON, which would lead to a PCE reduction. Therefore, M12 and M15 only need to be slightly higher to provide the required charging/discharging time to reduce the delay of the overall comparator. Table 1 summarizes the dimension values of the proposed rectifier circuit.

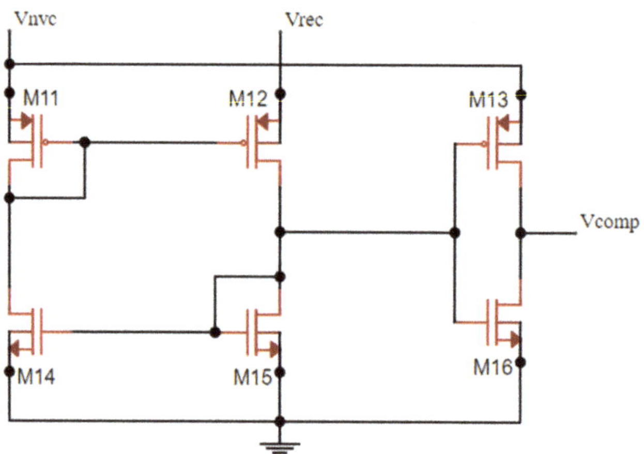

Figure 2. Schematic of the two-input common gate comparator CMP.

Table 1. Circuit transistor sizes.

	Unit Size (μm/μm)	Multiply Factor
M1/2/3/4	100/0.13	100
M5	100/0.13	50
M6/7/10/11/13/14/16	0.28/0.13	1
M7/9	20/0.13	1
M11/12	0.34/0.13	1

4. Results and Discussion

The simulation experiments were carried out using Cadence Virtuoso Analog Design Environment with a 130 nm CMOS process. The respective physical layout of the CMOS rectifier is presented in Figure 3. To replicate the output behavior of the energy harvester, the default input sinusoidal voltage amplitude and frequency used in the simulations were 600 mV and 3.2 kHz, respectively. Throughout most of the tests, C_{LOAD} and R_{LOAD}

were set at 2 μF and 5.5 kΩ to simulate the capacitance of the storing capacitor and the impedance of the electronics to be powered, respectively.

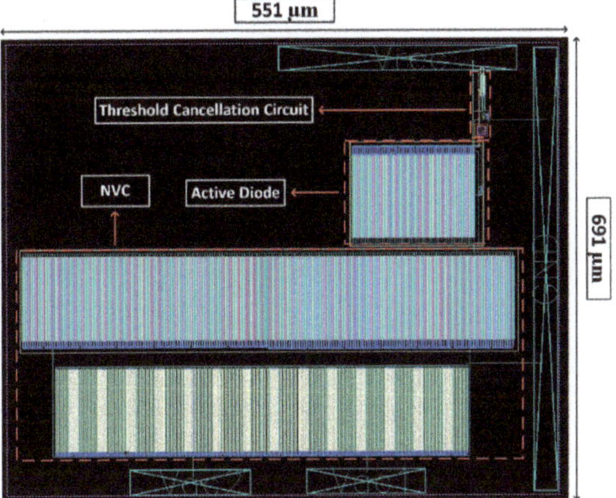

Figure 3. Physical layout of the proposed CMOS rectifier.

4.1. Transient Behavior

The transient performance of the output voltage, in both stages, is displayed in Figure 4. The first stage performs the full-wave rectification by converting the negative input voltages (V_{IN}) into positive ones (V_{NVC}). The voltage drop on this stage is around 1 mV, whereas the total voltage drop on the circuit is around 12 mV, which is possible due to the reduction of the internal resistance of the main pass transistor M5. The achieved voltage drop is crucial to enhance the output voltage across the load.

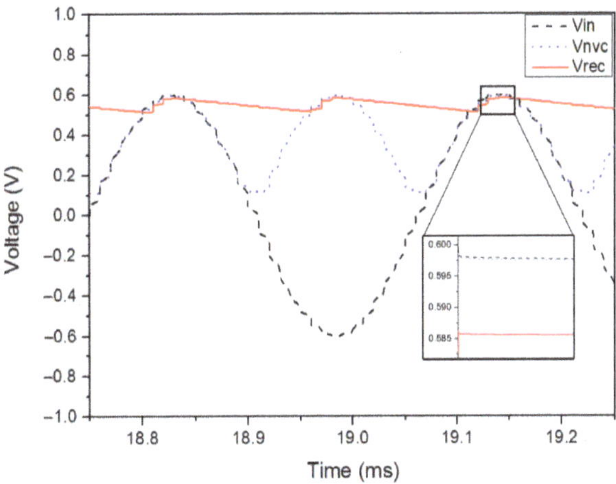

Figure 4. Simulated waveforms of the rectifier for $R_{LOAD} = 5.5$ kΩ and $C_{LOAD} = 2$ μF.

Figure 5 shows the VCE behavior versus the input voltage amplitude for different R_{LOAD} values. It is possible to observe that the proposed rectifier can work efficiently for an input voltage range from 0.45 V to 1 V for different ohmic loads, with a VCE varying between 96% and 99%. For an input voltage lower than 0.4 V, the VCE sharply decreases because the NVC transistors will enter the subthreshold region or even cut-off. Moreover, it can be noticed that the rectifier VCE is higher for larger load resistors, as would be expected.

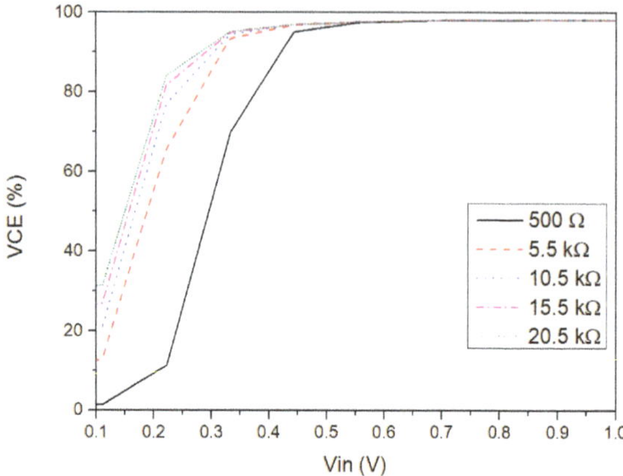

Figure 5. VCE versus input voltage amplitude simulated for different ohmic loads.

4.2. Reverse Leakage Current Analysis

The reverse leakage current analysis is one of the most important analyses to make in CMOS rectifiers because it affects the power efficiency of the overall system. This reverse leakage current is dependent on the delay of the comparator and, consequently, of the discharging path of the active diode provided by the AVC. Therefore, the analysis of the transient performance of the comparator is shown in Figure 6. It presents the output voltage of the comparator (V_{CMP}), the input and output voltage of the active diode used to perform the comparison, the gate voltage of M5 (V_{CAP}), and the current that flows through the active diode (I_{M5}). As can be observed, the comparator immediately turns on the gate of the AVC transistor to create the discharge path when V_{NVC} exceeds V_{rec}. At this stage, the current is flowing through M5, and V_{CAP} is low, which leads to a low voltage drop because V_{SG} is high. When V_{NVC} drops below V_{rec}, the comparator then quickly turns off the AVC, and consequently the active diode. Thus, the proposed structure does not exhibit reverse leakage current that would degrade the PCE of the proposed rectifier.

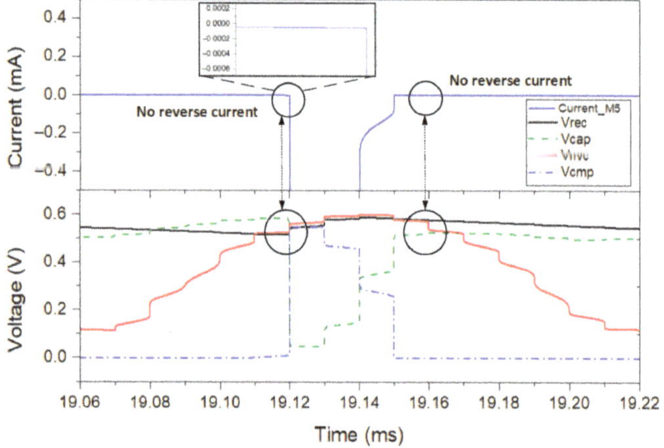

Figure 6. Simulated comparator behavior in steady state for $R_{LOAD} = 500\ \Omega$ and $C_{LOAD} = 2\ \mu F$.

4.3. Power Efficiency

The simulated power efficiency versus input voltage amplitude for different load resistors is presented in Figure 7. The definition of PCE is shown in (5):

$$PCE = \frac{\int_t^{t+T} V_{OUT}(t) \cdot I_{OUT}(t)\, dt}{\int_t^{t+T} V_{IN}(t) \cdot I_{IN}(t)\, dt} \cdot 100\%. \tag{5}$$

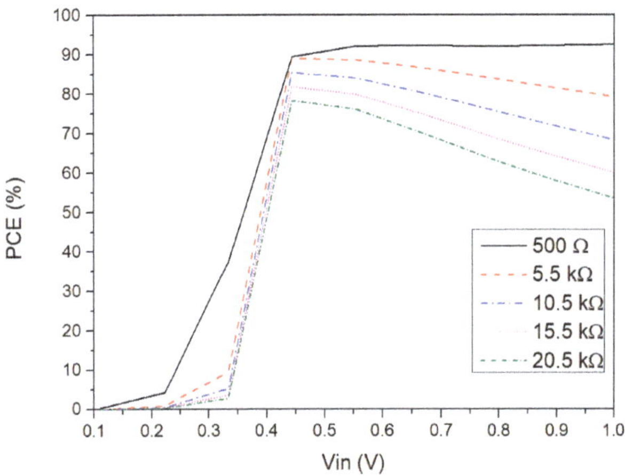

Figure 7. PCE versus input voltage amplitude simulated for different values of R_{LOAD}

The maximum PCE value of 94% can be found at 0.6 V for a R_{LOAD} of 500 Ω. When V_{in} is lower than this range, the PCE sharply decreases due to the low voltage efficiency, as noted in Section 4.1. Thus, the efficiency of the rectifier is poor in the ultra-low voltage range. Additionally, the PCE tends to decrease for higher input voltages because the power losses are mainly concentrated in the comparator. However, this case is not significant for ohmic loads lower than 15.5 kΩ. Moreover, for higher load resistors, the PCE tends to decrease due to the reduction of the output current, whereas the bias current that comes from the voltage source keeps almost constant. Regardless, from 0.45 V to 1 V, the power efficiency for low ohmic loads is considered as being good for this application. Additionally, the influence of the width of the NVC stage (M1–M4) and of M5 in both PCE and VCE can be observed in Figure 8. For this simulation test, the width of each stage was individually varied while the other was kept constant. This figure shows that the VCE and PCE features of both stages are at their maximum point for a width of 100 μm because the on-resistance of this transistor is directly influenced by the W/L ratio of the MOSFET. Even if the gate capacitance of M5 increases with the size, Figure 6 shows that the threshold cancellation circuit can drive this large transistor.

Figure 9 shows the power efficiency versus input voltage amplitude for different input frequencies. The load capacitor value was adapted to keep the output ripple voltage small depending on the input frequency. It is possible to observe that the proposed rectifier can achieve a high-power efficiency for low input frequencies in the operating voltage range. However, when the input voltage and frequency are high, the power efficiency tends to slightly decrease due to the power losses in the NVC and in the active diode, which in this case it is caused by the output signal of the comparator being too fast. Consequently, the working time of transistor M5 will be too short, which reduces the amount of power converted to the load. Nonetheless, at typical energy harvesting frequencies, the performance of the CMOS rectifier for the presented frequency range is suitable for this application.

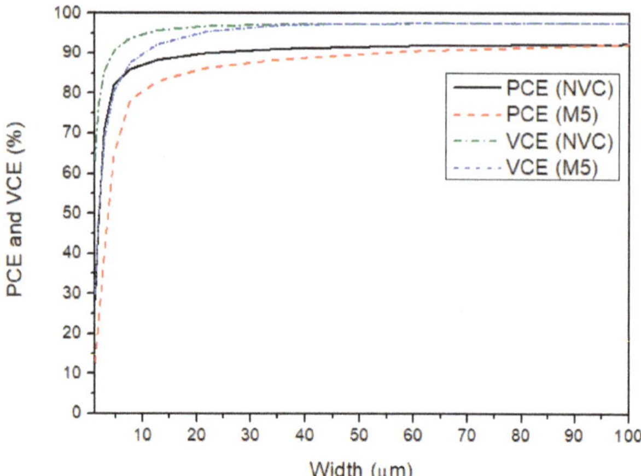

Figure 8. VCE and PCE features with the variation of the width of the NVC transistors and M5 (L = 0.13 μm).

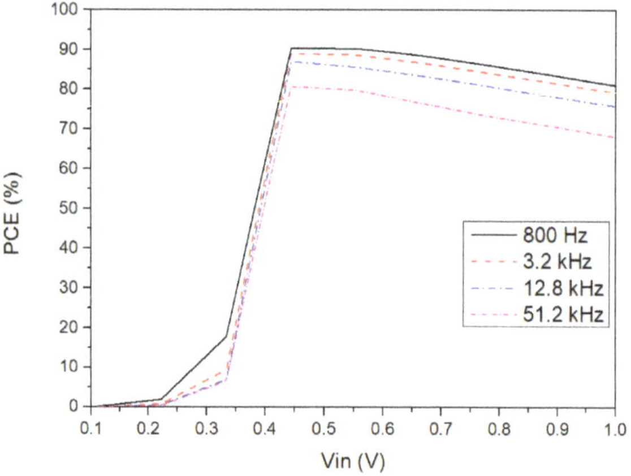

Figure 9. PCE versus input voltage amplitude simulated for different input frequencies for R_{LOAD} = 5.5 kΩ.

To prove the robustness of the proposed CMOS rectifier, it was tested through the four known process corners, such as the typical ones, fast, slow, slow NMOS, and fast PMOS, and fast NMOS and slow PMOS. Figure 10 presents the PCE plots with the variation of the temperature depending on the process corner. According to the simulation results, it can be observed that PCE tends to decrease when a fast PMOS is used due to the high speed of the active diode, which reduces the ON time of the rectifier. Consequently, the power transferred to the load is affected. Nevertheless, as long as the temperature rises, the power consumption of the rectifier also increases because the MOS threshold voltage is an exponential function of the temperature.

The performance comparison between this work and previous rectifiers is presented in Table 2. It shows that the proposed configuration can achieve higher VCE and PCE for a low voltage range. Even if the PCE in [29] is higher for a high ohmic load, for this application, it is only expected a low impedance of the electronics to be powered. Thus, the achieved VCE and PCE in this work are higher than those in the reported literature, highlighting its added value [15,28,30]. In addition, this work presents a wider input voltage range compared to the previously noted article. Therefore, it can be concluded that this rectifier can overcome the drawbacks of the structures discussed in Section 2, which means that this rectifier is very suitable for energy harvesting applications.

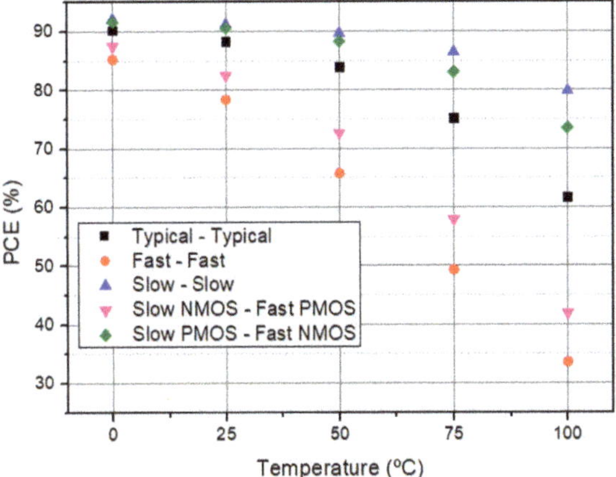

Figure 10. PCE variation with temperature depending on the process corner variation for a R_{LOAD} = 5.5 kΩ.

Table 2. Research comparison.

Ref.	Tech. (nm)	f_{in} (kHz)	V_{in} (V)	R_{LOAD} (kΩ)	C_{LOAD} (μF)	VCE (%)	PCE (%)
[15]	350	0.1	0.5	50	10	99	84
[28]	180	20	0.8	2	2	75	85
[29]	180	0.2	3	200	1	-	91.5
[30]	65	0.12	1.23	12	10	98	84
Proposed	130	3.2	0.45–1	5.5	2	99	80–90

5. Conclusions

A highly efficient active CMOS rectifier suitable to be applied to vibrational energy harvesters was presented in this work. The proposed structure was designed in 130 nm CMOS technology, and the results showed a VCE of 99% and a PCE of 80–90% for a low operation voltage from 0.45 V to 1 V and for an operating frequency of 3.2 kHz, which proves the value of this work for a practical energy harvesting application. These features were achieved by combining an NVC with an active diode biased by a threshold cancellation circuit, which dynamically reduces the threshold voltage effect. Moreover, this structure avoids the reverse leakage current due to the use of a no-delay comparator, which was vital to reduce the power losses.

The research work focused on developing a highly efficient rectifier to be integrated into a PMC. It is believed that this structure will efficiently contribute to solving the battery limitation problems of the WSNs for an environmental monitoring application. Further work should focus on integrating the proposed structure in the PMC and respective testing in real environmental conditions.

Author Contributions: A.G. was responsible for the investigation, design of the proposed circuit, simulations, respective validation, and the writing of the original draft. Z.Y. was involved in the investigation, simulations, respective validation, and in the writing revision and editing. T.D. supervised the research, administrated the project, and acquired the project funding. L.G. and P.M. were involved in the research supervision and validation of the simulation. Y.W. and P.L. were responsible for the conceptualization of the research and the administration of the project. Z.J. was also involved in the conceptualization of the research. All authors have read and agreed to the published version of the manuscript.

Funding: This work was supported by the National Natural Science Foundation of China under Project 61531008; by the "Praktisk og effektiv oppfølging av KOLS-pasienter i kommunene" from RFF Forskningsfond Oslofjordfondet under Project 285575; by the RFF Forskningsfond Agder under Project 321343; by the Chongqing Research Program of Basic and Frontier Technology under Project Cstc2018jcyjA3877, Project cstc2018jcyjAX0474, and Project cstc2019jcyj-msxmX0776; and by the Chongqing Education Commission–Science and Technology Research Program under Grant KJZD-K201800802, Grant KJZDK201900802, and Grant KLZD-K202000805.

Institutional Review Board Statement: Not applicable.

Informed Consent Statement: Not applicable.

Data Availability Statement: Not applicable.

Acknowledgments: Not applicable.

Conflicts of Interest: The authors declare no conflict of interest.

References

1. Adu-Manu, K.S.; Adam, N.; Tapparello, C.; Ayatollahi, H.; Heinzelman, W. Energy-harvesting wireless sensor networks (EH-WSNs) A review. *ACM Trans. Sens. Netw.* **2018**, *14*, 1–50. [CrossRef]

2. Dong, L.; Closson, A.B.; Jin, C.; Trase, I.; Chen, Z.; Zhang, J.X.J. Vibration-Energy-Harvesting System: Transduction Mechanisms, Frequency Tuning Techniques, and Biomechanical Applications. *Adv. Mater. Technol.* **2019**, *4*, 1900177. [CrossRef] [PubMed]

3. Iqbal, M.; Nauman, M.M.; Khan, F.U.; Abas, P.E.; Cheok, Q.; Iqbal, A.; Aissa, B. Vibration-based piezoelectric, electromagnetic, and hybrid energy harvesters for microsystems applications: A contributed review. *Int. J. Energy Res.* **2021**, *45*, 65–102. [CrossRef]

4. Toshiyoshi, H.; Ju, S.; Honma, H.; Ji, C.-H.; Fujita, H. MEMS vibrational energy harvesters. *Sci. Technol. Adv. Mater.* **2019**, *20*, 124–143. [CrossRef] [PubMed]

5. Guyomar, D.; Jayet, Y.; Petit, L.; Lefeuvre, E.; Monnier, T.; Richard, C.; Lallart, M. Synchronized switch harvesting applied to selfpowered smart systems: Piezoactive microgenerators for autonomous wireless transmitters. *Sens. Actuators A Phys.* **2007**, *138*, 151–160. [CrossRef]

6. Li, P.; Wen, Y.; Jia, C.; Li, X. A magnetoelectric composite energy harvester and power management circuit. *IEEE Trans. Ind. Electron.* **2010**, *58*, 2944–2951. [CrossRef]

7. Dell'Anna, F.; Dong, T.; Li, P.; Wen, Y.; Yang, Z.; Casu, M.R.; Azadmehr, M.; Berg, Y. State-of-the-Art Power Management Circuits for Piezoelectric Energy Harvesters. *IEEE Circ. Syst. Mag.* **2018**, *18*, 27–48. [CrossRef]

8. Li, Q.; Wang, J.; Ding, L.; Inoue, Y. A wide input amplitude range, highly efficient rectifier for low power energy harvesting systems. *Nonlinear Theory Its Appl. IEICE* **2014**, *5*, 499–511. [CrossRef]

9. Oh, T.; Islam, S.K.; Mahfouz, M.; To, G. A low-power CMOS piezoelectric transducer based energy harvesting circuit for wearable sensors for medical applications. *J. Low Power Electron. Appl.* **2017**, *7*, 33. [CrossRef]

10. Gomez-Casseres, E.A.; Arbulu, S.M.; Franco, R.J.; Contreras, R.; Martinez, J. Comparison of passive rectifier circuits for energy harvesting applications. In Proceedings of the 2016 IEEE Canadian Conference on Electrical and Computer Engineering (CCECE), Vancouver, BC, Canada, 15–18 May 2016. [CrossRef]

11. Panigrahi, A.; Sarania, S.; Brahma, R.G. A Low Voltage Rectifier for Piezo-Electric Energy Harvesting Designed in CMOS Technology. In Proceedings of the 2021 Devices for Integrated Circuit (DevIC), Kalyani, India, 19–20 May 2021; pp. 170–174. [CrossRef]

12. Farrarons, J.C.; Miribel Catala, P.; Saiz-Vela, A.; Vidal, M.P.; Samitier, J. Power-Conditioning Circuitry for a Self-Powered System Based on Micro PZT Generators in a 0.13-μm Low-Voltage Low-Power Technology. *IEEE Trans. Ind. Electron.* **2008**, *55*, 3249–3257. [CrossRef]

13. Yeo, K.H.; Ali, S.H.M.; Menon, P.S.; Islam, S. Comparison of CMOS rectifiers for micropower energy harvesters. In Proceedings of the 2015 IEEE Conference on Energy Conversion (CENCON), Johor Bahru, Malaysia, 19–20 October 2015; Volume 2015, pp. 419–423. [CrossRef]

14. Peters, C.; Spreemann, D.; Ortmanns, M.; Manoli, Y. A CMOS integrated voltage and power efficient AC/DC converter for energy harvesting applications. *J. Micromechanics Microengineering* **2008**, *18*, 104005. [CrossRef]

15. Peters, C.; Handwerker, J.; Maurath, D.; Manoli, Y. A Sub-500 mV highly efficient active rectifier for energy harvesting applications. *IEEE Trans. Circ. Syst. I Regul. Pap.* **2011**, *58*, 1542–1550. [CrossRef]

16. Niu, D.; Huang, Z.; Jiang, M. Paper A sub-0. *3V Highly Effic. CMOS* **2012**, *3*, 405–416.

17. Li, Q.; Wang, J.; Niu, D.; Inoue, Y. A Two-stage CMOS Integrated Highly Efficient Rectifier for Vibration Energy Harvesting Applications. *J. Int. Counc. Electr. Eng.* **2014**, *4*, 336–340. [CrossRef]

18. Fouad, H. Performance optimization of energy harvesting solutions for 0.18um CMOS circuits in embedded electronics design. *Cogent Eng.* **2020**, *7*, 1772947. [CrossRef]

19. Guo, S.; Lee, H. An Efficiency-Enhanced CMOS Rectifier With Unbalanced-Biased Comparators for Transcutaneous-Powered High-Current Implants. *IEEE J. Solid-State Circ.* **2009**, *44*, 1796–1804. [CrossRef]

20. Chang, R.C.-H.; Chen, W.-C.; Hsiao, W.-C. A high-performance AC-DC rectifier with fully actively controlled switches for vibration energy harvesting. In Proceedings of the 2017 IEEE Wireless Power Transfer Conference (WPTC), Taipei, Taiwan, 10–12 May 2017; pp. 12–15. [CrossRef]
21. Chang, R.C.-H.; Chen, W.-C.; Liu, L.; Cheng, S.-H. An AC–DC Rectifier With Active and Non-Overlapping Control for Piezoelectric Vibration Energy Harvesting. *IEEE Trans. Circ. Syst. II Express Briefs* **2019**, *67*, 969–973. [CrossRef]
22. Kotani, K.; Ito, T. High efficiency CMOS rectifier circuit with self-Vth-cancellation and power regulation functions for UHF RFIDs. In Proceedings of the 2007 IEEE Asian Solid-State Circuits Conference, Jeju, Korea, 12–14 November 2007; pp. 119–122. [CrossRef]
23. Kotani, K.; Ito, T. Self-Vth-cancellation high-efficiency CMOS rectifier circuit for UHF RFIDs. *IEICE Trans. Electron.* **2009**, *E92-C*, 153–160. [CrossRef]
24. Khan, S.; Choi, G. High-efficiency CMOS rectifier with minimized leakage and threshold cancellation features for low power bio-implants. *Microelectron. J.* **2017**, *66*, 67–75. [CrossRef]
25. Hashemi, S.S.; Sawan, M.; Savaria, Y. A high-efficiency low-voltage CMOS rectifier for harvesting energy in implantable devices. *IEEE Trans. Biomed. Circ. Syst.* **2012**, *6*, 326–335. [CrossRef]
26. Roy, K.; Mukhopadhyay, S.; Mahmoodi, H. Leakage current mechanisms and leakage reduction techniques in deep-submicrometer CMOS circuits. *Proc. IEEE* **2003**, *91*, 305–327. [CrossRef]
27. Lee, S.-Y.; Liao, Z.-X.; Lee, C.-H. Energy-harvesting circuits with a high-efficiency rectifier and a low temperature coefficient bandgap voltage reference. *IEEE Trans. Very Large Scale Integr. (VLSI) Syst.* **2019**, *27*, 1760–1767. [CrossRef]
28. Lakshmi, N.V.; Vinoth, G.; Binoj, J.; Manikandan, N. Design of high efficiency energy harvesting circuit using dual switching technique. *Mater. Today Proc.* **2021**, *39*, 725–730. [CrossRef]
29. Din, A.U.; Kamran, M.; Mahmood, W.; Aurangzeb, K.; Altamrah, A.S.; Lee, J.-W. An efficient CMOS dual switch rectifier for piezoelectric energy-harvesting circuits. *Electronics* **2019**, *8*, 66. [CrossRef]
30. Yuen, P.W.; Chong, G.C.; Ramiah, H. A high efficient dual-output rectifier for piezoelectric energy harvesting. *AEU—Int. J. Electron. Commun.* **2019**, *111*, 152922. [CrossRef]

Article

Concurrent-Mode CMOS Detector IC for Sub-Terahertz Imaging System

Moon-Jeong Lee [1], Ha-Neul Lee [1], Ga-Eun Lee [1], Seong-Tae Han [2] and Jong-Ryul Yang [1,*]

[1] Department of Electronic Engineering, Yeungnam University, Gyeongsan 38541, Korea; hsg03170@yu.ac.kr (M.-J.L.); gksmf4461@yu.ac.kr (H.-N.L.); gaeun@ynu.ac.kr (G.-E.L.)

[2] Electrophysics Research Center, Korea Electrotechnology Research Institute, Changwon 51543, Korea; saiph@keri.re.kr

* Correspondence: jryang@yu.ac.kr; Tel.: +82-53-810-2495

Abstract: A CMOS detector with a concurrent mode for high-quality images in the sub-terahertz region has been proposed. The detector improves output-signal coupling characteristics at the output node. A cross-coupling capacitor is added to isolate the DC bias between the drain and gate. The detector is designed to combine a 180° phase shift based on common source operation and an in-phase output signal based on the drain input. The circuit layout and phase shift occurring in the cross-coupled capacitor during phase coupling are verified using an EM simulation. The detector is fabricated using the TSMC 0.25-μm mixed-signal 1-poly 5-metal layer CMOS process, where the size, including the pad, is 1.13 mm × 0.74 mm. The detector IC comprises a folded dipole antenna, the proposed detector, a preamplifier, and a voltage buffer. Measurement results using a 200-GHz gyrotron source demonstrate that the proposed detector voltage responsivity is 14.13 MV/W with a noise-equivalent power of 34.42 pW/$\sqrt{\text{Hz}}$. The high detection performance helps resolve the 2-mm line width. The proposed detector exhibits a signal-to-noise ratio of 49 dB with regard to the THz imaging performance, which is 9 dB higher than that of the previous CMOS detector core circuits with gate-drain capacitors.

Keywords: CMOS detector; concurrent-mode; differential detector IC; imaging SNR; integrated folded-dipole antenna; sub-terahertz imaging; voltage responsivity

Citation: Lee, M.-J.; Lee, H.-N.; Lee, G.-E.; Han, S.-T.; Yang, J.-R. Concurrent-Mode CMOS Detector IC for Sub-Terahertz Imaging System. *Sensors* **2022**, 22, 1753. https://doi.org/10.3390/s22051753

Academic Editor: Roman Sotner

Received: 7 January 2022
Accepted: 22 February 2022
Published: 23 February 2022

Publisher's Note: MDPI stays neutral with regard to jurisdictional claims in published maps and institutional affiliations.

1. Introduction

Terahertz (THz) waves represent frequencies of the 0.1–10 THz bands of the frequency spectrum. Located between radio and light waves, THz waves with high transmittance and directivity are expected to be widely used in the fields of security imaging, radio astronomy, medical imaging, and art [1–7]. A THz imaging system is capable of nondestructive inspection owing to the low ionization energy with regard to frequency. In recent times, the THz wave has been of interest as a frequency resource to replace the millimeter wave, as it can be used in communication beyond the limit of short-wavelength waves [8].

Unlike millimeter wave infrastructure, an active full-wave structure is required to develop a safe and high-resolution THz imaging system. The performance of an active system can be expressed by the signal-to-noise ratio (SNR), in which a measurement target is independently placed between a transmitter and a receiver, and the receiver detects and outputs the signal reflected or transmitted from the target. The SNR of the image quality factor can be calculated as the ratio between the output signals when the transmitted signal is reflected away by a metal target and when the transmitted signal passes through a nonreflective target; the reflected transmission output measures the noise of the detector. The image quality of the output can be measured using high-power sources; however, the implementation of the high-powered transmitter at the THz frequency band is difficult and expensive to fabricate. A receiver with high sensitivity characteristics improves the performance of an imaging system. The sensitivity of the receiver is measured based on

81

voltage responsivity (R_V) and noise-equivalent power (*NEP*). R_V represents the output voltage magnitude as a function of the input signal power, and *NEP* demonstrates the noise characteristics of the detector. A receiver with low *NEP* exhibits a smaller power threshold to distinguish between signal and noise, allowing for more sensitive operation. High R_V detectors can minimize *NEP*, resulting in high performance [9].

The surface plasma phenomenon of field-effect transistors, which was first introduced by M. Dyakonov and M. Shur, exhibited potential for sensing in the THz band using a standard complementary metal-oxide-semiconductor (CMOS) process [10,11]. A CMOS detector capable of detecting a signal above the operating frequency is known as a self-mixing detector or a square row detector, and outputs a direct current (DC) voltage [12]. Body bias control can modify the electrical characteristics of the detector core to improve the performance by changing the subthreshold slope [13]. In addition, a CMOS detector with body bias control has demonstrated a wide dynamic range with strong voltage responsivity [13]. Detector bias changes the electric field effect of the detector, which requires various operating parameters. Cascode topology allows a detector to operate similar to a general amplifier, where the source, gate, and drain interfaces are simplified, thereby significantly improving the detector performance [14–18]. CMOS detector topologies have been proposed with commonly used gate and drain input structures, where the phase is determined by adding transconductance terms using Taylor expansion [19]. The analysis confirmed that detection performance can be improved based on the differential phase input with drain bias. A detector circuit with weak inversion mode operation characteristics can help improve performance through the phase coupling of the output signal and unbiased drain.

In this study, a concurrent-mode CMOS detector integrated circuit (IC) with phase-coupled operation at the output node with input signals routing through cross-coupled capacitors is proposed. The proposed detector IC comprises a differential-folded dipole antenna, the proposed differential detector core, a pre-amplifier to convert a signal from a differential to single-ended mode, and a voltage buffer with low gain. The performance of the detector can be improved by combining the two signals of the dual detection output in one core circuit; as a result, the proposed IC functions as a high-quality THz imaging system. Section 2 describes the core configuration and operating principle of the CMOS concurrent-mode detector with phase combination. A phase difference occurs between the gate of the input node and the drain of the output node owing to the cross-coupled capacitors. The proposed detector IC was implemented using the TSMC 0.25-µm CMOS 1-poly 5-metal (1P5M) process. Section 3 presents the measurement results of the performance of the detector IC and results obtained using 200-GHz raster-scanned imaging. Section 4 concludes the study.

2. Proposed Detector Circuit

The detector circuit with a gate-drain capacitor used to enhance the potential difference between the drain and source terminals is shown in Figure 1a. Extra gate-drain capacitors are added to a CMOS-based single gate input circuit and exhibit a high potential difference. Despite its high performance, the previous topology is sensitive to the capacitor size, making it difficult to guarantee operational safety. The limitations of the process model and layout used to optimize the capacitor size can prevent the detector from achieving optimal performance. A high-performance concurrent-mode detector circuit with safe operation via cross-coupled structure for dual operation has been proposed.

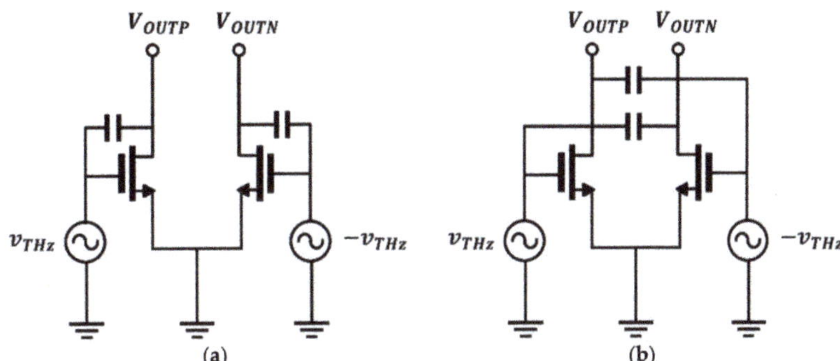

Figure 1. CMOS detector configuration: (**a**) Previous detector circuit with gate-drain capacitors; (**b**) proposed concurrent-mode detector.

2.1. Proposed Detector Core Configuration

The proposed detector with a concurrent-mode operation circuit, as shown in Figure 1b, improves the quality of the THz imaging system; the circuit operates based on the square root detection of the input power and outputs a voltage signal. The proposed detector consists of the same components as the detector circuit with gate-drain capacitors, except for cross-coupled capacitors. The cross-coupled capacitors were used to transmit the THz signal to the drain of an adjacent detection transistor. Unlike the gate-drain capacitor circuit, which improves the drain-source potential difference, the capacitor blocks the gate input DC bias and only transmits the in-phase signal. The detector structure can be interpreted as a combination of the gate input and drain input circuits. The proposed detector has two operation modes. One of the operation modes of the proposed detector can be analyzed in the same method as the general common-source stage amplifier shown in Figure 2a. The output signal is inverted from the input signal, with a 180° difference.

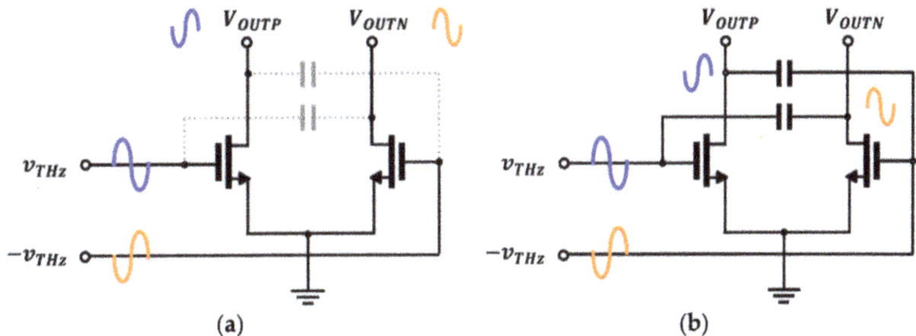

Figure 2. Two operations in the proposed circuit: (**a**) gate input topology; (**b**) drain input topology.

Figure 2b shows drain input detection operation using cross capacitors connected to the drain nodes of the two transistors. The output signal of the drain input circuit exhibits the same phase as the gate input without any phase shift and is, therefore, phase-coupled to the output drain terminal. Both detector inputs include gate bias for weak inversion operation at the gate node. The operating principle of the proposed detector is defined as the concurrent-mode operation, in which an incident signal is simultaneously applied to the gate and drain of the single detector core. The concurrent-mode operating characteristics are designed using cross-coupling capacitors; a sufficient capacitance is required for high-frequency coupling. The same phase at the final output node is important for combining the two detector outputs. The proposed detector circuit was designed to incorporate the cross-coupled capacitor layout and additional core circuit phase shifts and was validated using electromagnetic simulation performed using the Keysight Advanced Design System software.

The performance of the two detector cores, as shown in Figure 1, was simulated using Cadence Spectre. Although achieving the impedance-matching condition is essential for optimizing detector characteristics, two detector cores were used as transistors of the same size to compare the performance based on the detector configuration. Figure 3 shows the simulated voltage responsivities between the previous and proposed detector cores. The proposed concurrent-mode detector core exhibited a voltage responsivity of 1.5–3.3 times higher than that of the previous detector core based on the input power. The performance improvement owing to phase coupling is confirmed above the power level at which the two detection operation outputs are significant. However, beyond a certain power level, a sufficiently large power signal is incident on the gate node, and the output of the gate input circuit becomes dominant, decreasing the performance difference.

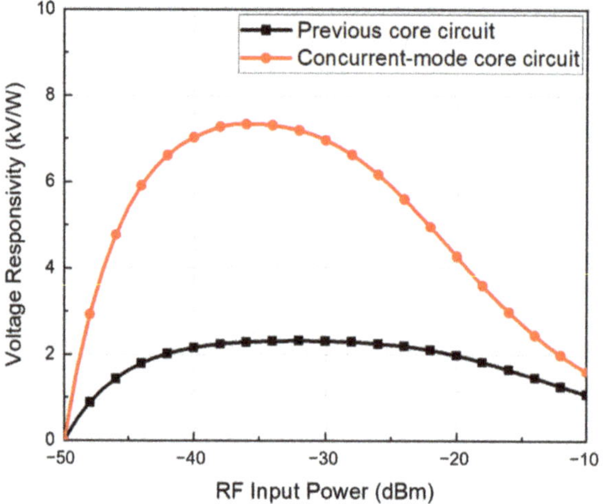

Figure 3. Simulated voltage responsivity of the detector core with gate-drain capacitors and the proposed concurrent-mode detector core.

2.2. Folded Dipole Antenna

A non-frequency selective detection circuit is determined for operating frequencies using an integrated antenna. A differential integrated antenna with an operating frequency of 200 GHz was designed to obtain input signals whose frequencies are higher than the device operating frequency. A grounded guard ring was placed at a sufficient distance from the antenna metal to focus on the internal electric field. The total area of the antenna, including the guard ring, is 500 μm × 200 μm. Compared with a patch antenna, the proposed antenna has a smaller area and exhibits similar performance, which is advantageous for a large-scale array. The on-chip antenna was configured as a folded dipole antenna to assume the operational mode of the detector circuit by applying a gate bias through a virtual ground. The radiating metal was folded at 45° to balance the paths of the inner and outer lengths. Chamfered radiating metal edges can prevent distortion from the antenna owing to processing changes. The simulation results using the 3-D EM simulation tool, ANSYS Electronics, demonstrate 11 GHz of −10 dB bandwidth corresponding to 195–206 GHz, as depicted in Figure 4a. The simulation data in Figure 4b represent the E-field and H-field characteristics in the far field. As shown in Figure 4c, the antenna radiation gain is simulated at −2.79 dBi at 200 GHz, exhibiting a peak radiation efficiency of 90.5%. In the far field, the simulated data confirm that the antenna performance is omnidirectional and suitable for image measurement.

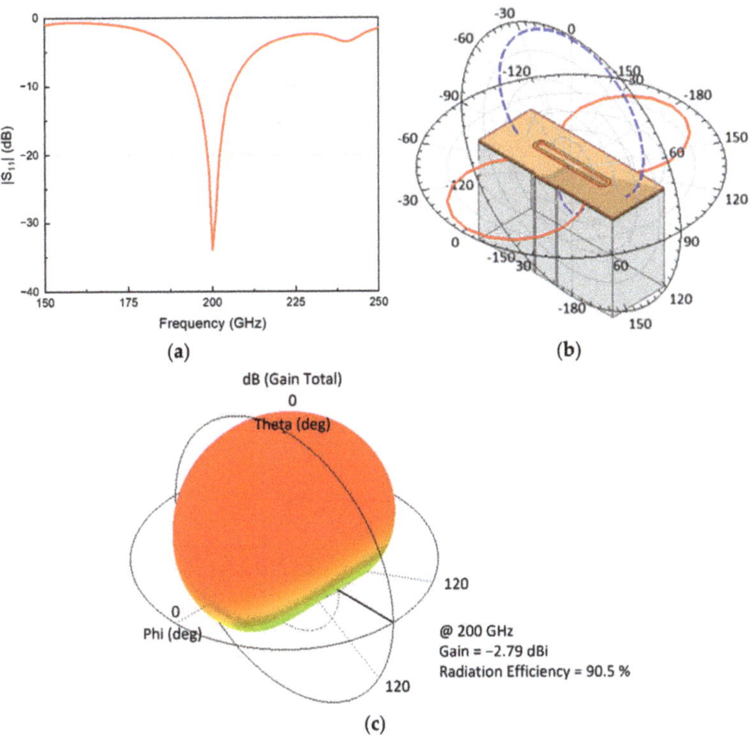

Figure 4. Simulation results of the integrated folded dipole antenna: (**a**) reflection coefficient $|S_{11}|$; (**b**) far-field radiation pattern at 200 GHz; (**c**) 3D radiation pattern at 200 GHz.

2.3. CMOS Detector IC Implementation

The CMOS detector IC, including the in-phase coupled detector, is illustrated in Figure 5. The final output signal was generated using a differential-to-single-ended preamplifier and monitored using an impedance-converting voltage buffer. A validated preamplifier and voltage buffer used the same circuit to compare the inherent detector core circuits [19].

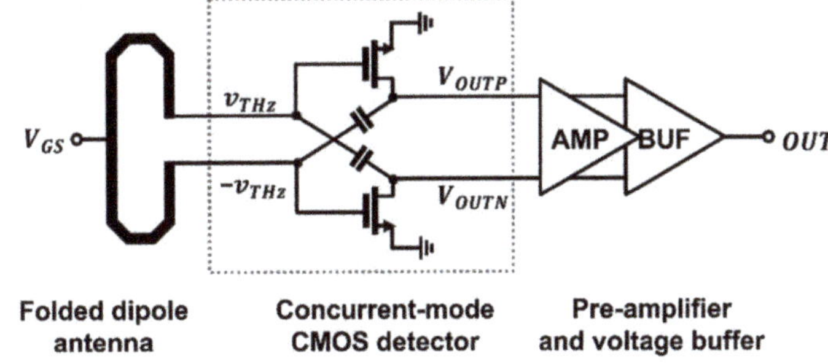

Figure 5. CMOS detector IC comprising a 200-GHz folded dipole antenna, the proposed detector, a differential-to-single-ended preamplifier, and a voltage buffer.

Figure 6 shows the circuit used to monitor the output. Voltages V_{B1}, V_{B2}, V_{B3}, and V_{B4} are biased in the circuit based on a current reference circuit. A common gate transistor, M2, acting as a level shifter and isolator, was used to change the output voltage of the transconductance stage at the load. The output signals of the detector core were transferred to transistors M3 and M4, which operated in the subthreshold region by self-biasing to the DC output voltage of the core. The converted DC and coupled sub-terahertz signals at terminals V_{OUTP} and V_{OUTN} are combined as currents at the drain node of M3 and M4. In the preamplifier, the in-phase DC signals are summed and the out-of-phase sub-terahertz signals are canceled [2]. In this operation, the differential input is converted into a single-ended output at the detector IC. The output of the preamplifier with unity gain is connected to M7 of the source follower via negative feedback for operational stability to output the final signal.

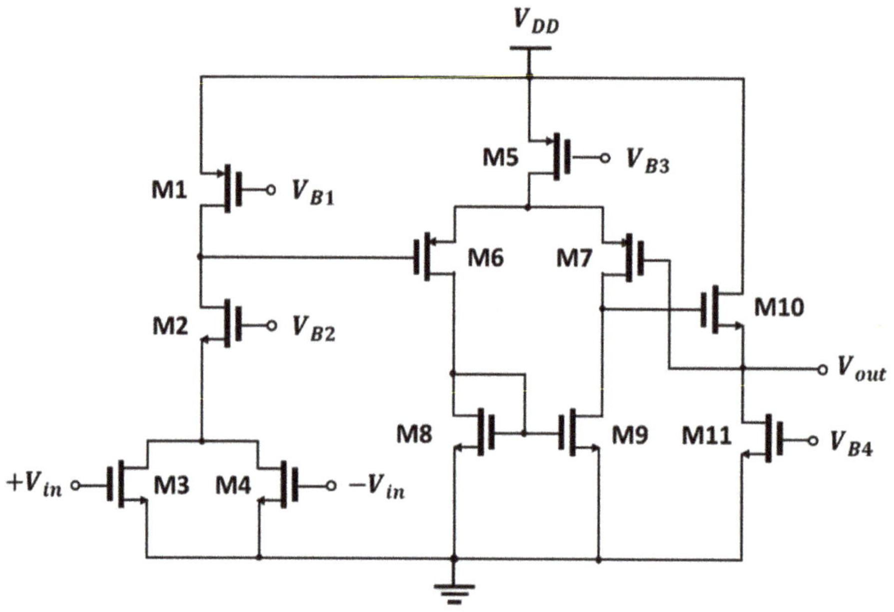

Figure 6. Schematic of the preamplifier with a voltage buffer.

The proposed detector IC is implemented using the TSMC 0.25-μm mixed-signal process, as shown in Figure 7. The gate bias of the detector core is applied to the alternating current (AC) ground node of the folded dipole antenna using an external instrument. The additional isolation was provided by the integrated resistance of 60 kΩ between the node and an I/O pad. Bias voltages except the gate voltage were provided by integrated low-dropout (LDO) regulator and current reference (I_{REF}) circuits to ensure the operational safety of the proposed detector. The size of the fabricated chip including the pad and proposed detector circuit is 1.13 mm × 0.74 mm. The difference between the two differential output DC voltages and bias voltage offset each other, and signal leakage into the output signal is attenuated by a radio-frequency (RF) choke at the input transistors of the preamplifier; consequently, the differential input is converted to a single-ended output. The detected signal is delivered to the final output pad using a voltage buffer with a gain of −1.5 dB, which determines the detector performance based on impedance conversion.

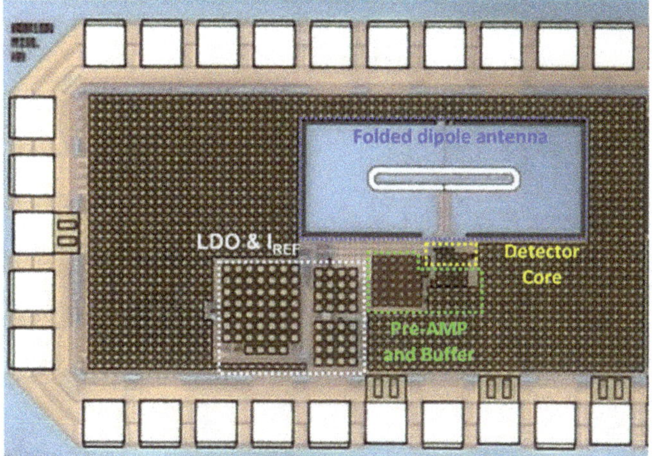

Figure 7. Die photograph of the proposed detector IC implemented using the TSMC 0.25-µm mixed-signal CMOS process comprising a 200-GHz folded dipole antenna, the proposed detector core, a pre-amplifier, and a voltage buffer.

3. Measurement Results and Discussions

3.1. Measurement Setup

The setup used to measure the performance of the proposed detector is illustrated in Figure 8. The 200-GHz signal generated by the gyrotron source demonstrates Gaussian beam characteristics, whose focal plane is aligned by off-axis parabolic (OAP) mirrors [20] with a focal length of 15.24 cm. A physical chopper located in the focal plane reduces the flicker noise by transmitting the detected output DC voltage along with an AC signal. The measurements were conducted with a modulation frequency of 200 Hz for avoiding the effect of switching noise from the power supply. To monitor the constant voltage output of the LDO regulator, digital 4-bit control signals were applied using the National Instruments (NI) data acquisition board (DAQ). Before reaching the monitoring equipment, the detector IC output was amplified with a gain of 5 through a bandpass filter of 100–300 Hz using an SR560 low-noise voltage amplifier manufactured by Stanford Research Systems. R_V was measured using an oscilloscope, and *NEP* was measured using the Keysight N9010B signal analyzer. The unamplified detector performance was measured by dividing the measurement with the amplifier gain.

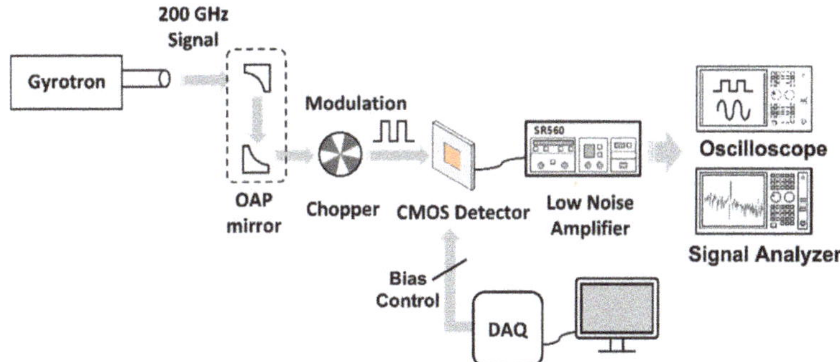

Figure 8. Measurement setup to analyze the performance of the proposed detector IC. An oscilloscope to measure R_V and a signal analyzer to obtain noise-equivalent power have been used.

In the THz imaging system test, as shown in Figure 9a, the distance between the mirror and the sample was 420 mm, and that between the sample and the proposed CMOS detector IC was 40 mm; the sample was placed on the XY stage. During image acquisition, the DAQ generated a digital 4-bit control code and analyzed the measurement data. The output signals were acquired using the NI DAQ hardware and NI LabVIEW software tools. The sample was moved at intervals of 1 mm in the measurement environment, as shown in Figure 9b, and the final output image was obtained using 2-D raster scanning.

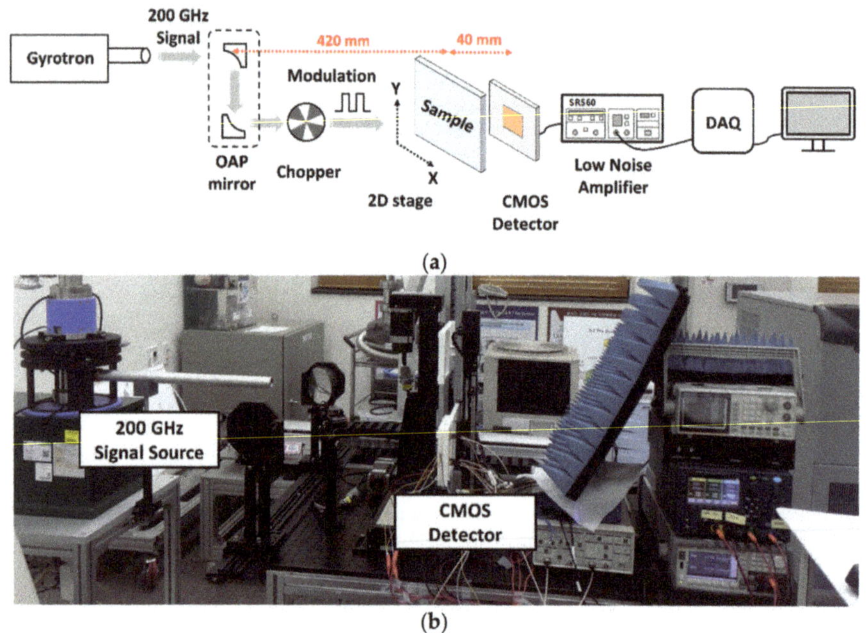

Figure 9. Measurement setup: (**a**) block diagram for THz imaging using raster scanning; (**b**) experimental setup.

3.2. Proposed Detector IC Performance

The performance of sub-THz CMOS detectors is determined by R_V and NEP. R_V is defined as the change in the output voltage based on the presence or absence of the incident signal with the specific power applied to the detector; it is calculated as:

$$R_V = \frac{V_{OUT} - V_{DCOFF}}{P_{IN}} = \frac{V_{OUT} - V_{DCOFF}}{P_D \cdot A_{EFF}} \ [V/W], \tag{1}$$

where V_{OUT} is the output voltage when the input power P_{IN} is applied to the detector, V_{DCOFF} is the output voltage without the incident signal, P_D is the power density incident to the detector IC, and A_{EFF} is the effective antenna area, which includes the integrated antenna and wavelength characteristics [9,13]. NEP is defined as the input power level that becomes equal to the noise generated from the detector itself; it is expressed using R_V as:

$$NEP = \frac{\sqrt{N_V}}{R_V} \ \left[W/\sqrt{Hz}\right], \tag{2}$$

where N_V denotes the noise spectral density [9].

When using a receiver antenna to transmit a THz signal to a detector, it is vital to determine the characteristics of the detector antenna and consider the input power equation of the detector to accurately analyze and measure the detector performance. The unit power density of the gyrotron measured at the detector position is 0.5 W/m^2, and the effective area considering the receiving antenna gain at 200 GHz is 9.62×10^{-8} m^2 [13]. The input power was calculated based on the measured performance, considering the difference in antenna gain according to the radiation area and similar antenna simulation values [21]. Figure 10 shows the measured R_V and *NEP* values of the proposed CMOS detector with different gate bias voltages. The results exhibit high R_V and low noise when the gate bias is lower than the threshold voltage. R_V and *NEP* were 14.13 MV/W and 34.42 pW/$\sqrt{\text{Hz}}$, respectively, under the gate bias condition of 150 mV. Table 1 lists the performance comparison of the proposed CMOS detector with previously developed detector core configurations. The proposed CMOS detector exhibits higher R_V compared with the other detectors. As a result of comparing detectors with the same minimum gate length, the proposed detector showed the lowest *NEP* and highest R_V. In previous studies, while calculating the effective area of an antenna, the difference in the radiation area between the antenna simulations and measurements was not considered. The performance of the proposed detector was calculated using the simulation data of the integrated antenna, which includes the ground guard ring in the simulation model for providing the same radiation area as the fabricated IC.

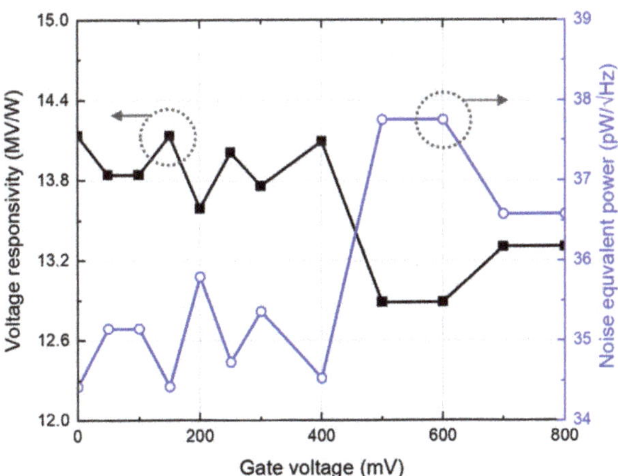

Figure 10. Measurement results of the voltage responsivity and the noise-equivalent power using the proposed CMOS detector IC at 200 GHz.

Table 1. Comparisons of CMOS detector performance.

Ref.	Process (nm)	Freq. (GHz)	Detector Core Configuration	R_V (kV/W)	*NEP* (pW/$\sqrt{\text{Hz}}$)
[13]	250	200	Gate-drain cap.	5696	62.4
[16]	65	310	Drain input	2	3.5 [1]
[19]	250	200	Gate-drain cap.	357.1	57.3
[22]	250	200	Gate-drain cap.	2020	76
[23]	90	365	Gate source input	1200	200
[24]	250	200	Gate-drain cap.	2990	46.3
This works	250	200	Concurrent-mode	14,130	34.42

[1] Measured in a Faraday cage.

3.3. Images Obtained Using the Proposed Detector IC

Copper foil tapes of different thicknesses were placed on the Styrofoam substrate to measure the resolution of the proposed detector, as shown in Figure 11a. The sample size was 50 mm × 50 mm. Considering that the wavelength of 200 GHz in the air is 1.5 mm, the sample was manufactured considering a thickness of ≥2 mm. The real sample was digitized, as shown in Figure 11b, for a digital area comparison using MATLAB. The sub-THz imaging at 200 GHz yielded results that are 63.6% identical to those of the digitalized sample. As shown in Figure 12, imaging results obtained using the proposed detector demonstrated that a 2-mm thick conductive target could be distinguished from the background. The measurement image is more distributed than the physical sample, as the passing waves are dispersed over the distance of 40 mm between the detector and sample. Considering the difficulty in identifying the wavelength width of the metal using a CMOS detector owing to the distance between the sample and detector, the measurement result exhibits a high-resolution image. All the images were compared by normalizing them to either maximum or minimum ratios.

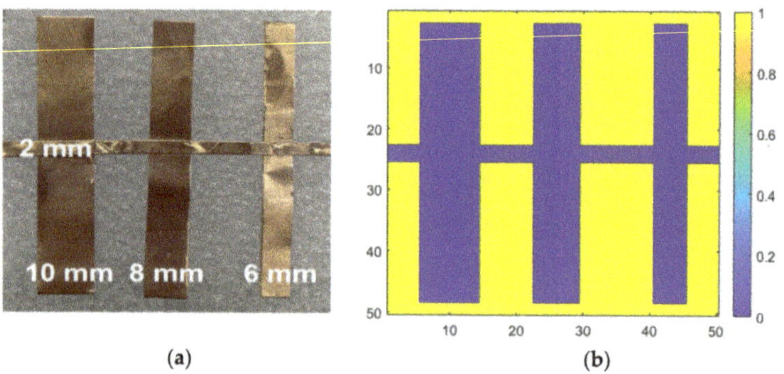

(a) (b)

Figure 11. Sample target for sub-THz imaging: (**a**) photograph of a sample with different copper widths; (**b**) digitalized image of the sample to image correlation.

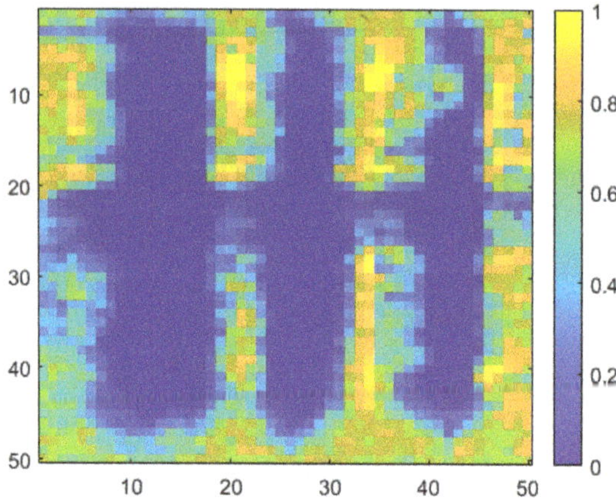

Figure 12. Measurement image using the proposed CMOS detector IC.

As illustrated in Figure 13a, the sample is used to compare the effect of the difference in the detector core circuit, and the individual detectors are considered under the optimal detection performance conditions [23]. Figure 13b shows the image obtained using the model capacitor, which is supported by the process design kit (PDK). Figure 13c shows an image obtained from a detector designed to achieve optimal detection performance using a customized capacitor in the same core circuit structure. The proposed detector, as shown in Figure 14, exhibited a 59.37% match with the normalized image, whereas previous studies demonstrated correlations of 41.4% and 53.7%. The concurrent-mode detector containing the cross-coupled capacitors better resolves the inner plus-shaped copper foil, which was impossible to identify in previous studies. The image results show that the image SNR of a single frame is 49 dB, which is 9 dB higher than that of the image obtained using the previous detector in one frame.

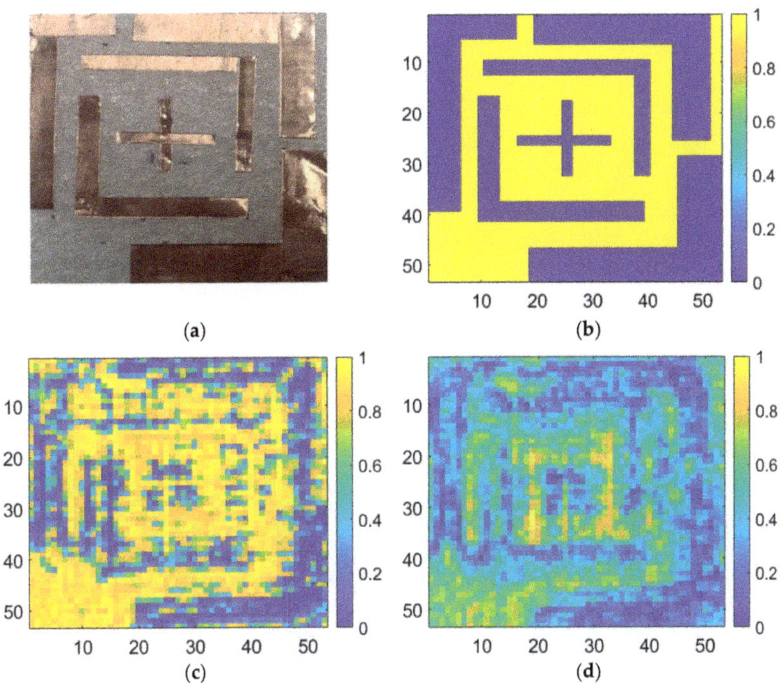

Figure 13. Sub-THz imaging in the previous study [23]. (**a**) Photograph of a real sample; (**b**) digitized image; (**c**) using a common-source detector circuit with the standard capacitors in the process design kit; (**d**) using a common-source detector circuit with the customized capacitors.

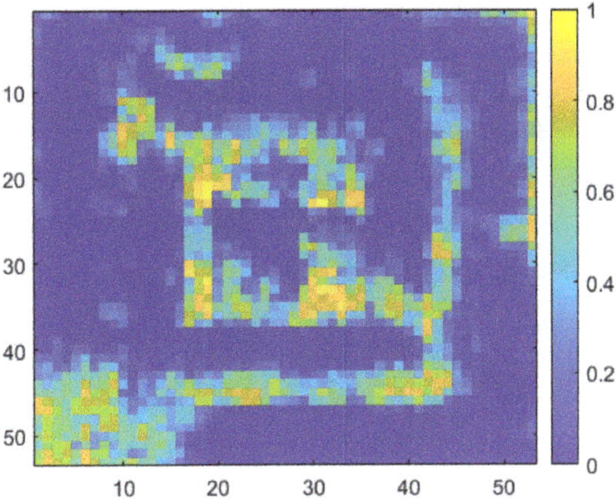

Figure 14. Imaging measurement using the proposed concurrent-mode detector at 200 GHz.

4. Conclusions

A CMOS detector with concurrent in-phase coupling was proposed to achieve high-quality images using THz imaging systems. The cross-capacitor structure possessed two detecting operations in a common source and a drain input structure while considering general amplifier analysis. The simulation results demonstrated that the proposed detector exhibits higher detection performance than the previously studied detector topology using gate-drain capacitors. At the output stage, the detector performance with regard to phase coupling was improved by 1.5–3 times higher than that of the previous detector core based on the input power. The detector, manufactured by a TSMC 0.25 µm CMOS process, comprised a differential folded dipole antenna, as the proposed core was connected by cross-coupled capacitors, a pre-amplifier, and a low-gain voltage buffer amplifier. The values of R_V and *NEP* at 200 GHz were 14.13 MV/W and 34.42 pW/$\sqrt{\text{Hz}}$, respectively, at a gate bias of 150 mV. In contrast to the previous detector studies, the proposed detector structure has a smaller detector IC area with higher detection performance. At 200 GHz, the measurements of a THz imaging system using samples of copper foil tape attached to Styrofoam substrates demonstrated that the proposed detector can resolve wavelengths (approximately 200 GHz) of 2-mm thickness with a high correlation coefficient. The proposed detector demonstrated an improved correlation of 59.37% with the actual sample, 1.4 times higher than the previous detector under identical conditions, except for the circuit structure. The image SNR, which indicates the image quality, was 49 dB, which was 9 dB higher than that obtained using the model capacitor of the process. The THz image quality was improved using the proposed concurrent-mode CMOS detector without the need for an additional circuit.

Author Contributions: Conceptualization, J.-R.Y.; methodology, M.-J.L., G.-E.L. and J.-R.Y.; software, M.-J.L. and H.-N.L.; validation, M.-J.L. and J.-R.Y.; formal analysis, M.-J.L. and J.-R.Y.; investigation, M.-J.L.; resources, S.-T.H. and J.-R.Y.; data curation, M.-J.L. and J.-R.Y.; writing—original draft preparation, M.-J.L., H.-N.L., G.-E.L. and J.-R.Y.; writing—review and editing, M.-J.L., S.-T.H. and J.-R.Y.; visualization, M.-J.L.; supervision, J.-R.Y.; project administration, J.-R.Y. All authors have read and agreed to the published version of the manuscript.

Funding: This work was partly supported by the National Research Foundation of Korea grant funded by the Korea government (MSIT) (No. 2021R1A2C2004356) and Institute of Information & Communications Technology Planning & Evaluation grant funded by the Korea government (MSIT) (No. 2018-0-00711).

Institutional Review Board Statement: Not applicable.

Informed Consent Statement: Not applicable.

Data Availability Statement: The data presented in this study are available on request from the corresponding author. The data are not publicly available due to the regulation of the project.

Acknowledgments: The chip fabrication and EDA tool usage were partly supported by the IC Design Education Center (IDEC), Korea.

Conflicts of Interest: The authors declare no conflict of interest.

References

1. Hu, B.B.; Nuss, M.C. Imaging with terahertz waves. *Opt. Lett.* **1995**, *20*, 1716–1718. [CrossRef] [PubMed]
2. Hillger, P.; Grzyb, J.; Jain, R.; Pfeiffer, U.R. Terahertz imaging and sensing applications with silicon-based technologies. *IEEE Trans. Terahertz Sci. Technol.* **2019**, *9*, 1–19. [CrossRef]
3. Kim, J.; Yoon, D.; Yun, J.; Song, K.; Kaynak, M.; Tillack, B.; Rieh, J.-S. Three-Dimensional Terahertz Tomography with Transistor-Based Signal Source and Detector Circuits Operating Near 300 GHz. *IEEE Trans. Terahertz Sci. Technol.* **2018**, *8*, 482–491. [CrossRef]
4. Yun, J.; Oh, S.J.; Song, K.; Yoon, D.; Son, H.Y.; Choi, Y.; Huh, Y.-M.; Rieh, J.-S. Terahertz reflection-mode biological imaging based on InP HBT source and detector. *IEEE Trans. Terahertz Sci. Technol.* **2017**, *7*, 274–283. [CrossRef]
5. Kawase, K.; Ogawa, Y.; Watanabe, Y.; Inoue, H. Non-destructive terahertz imaging of illicit drugs using spectral fingerprints. *Opt. Express* **2003**, *11*, 2549–2554. [CrossRef] [PubMed]
6. Liu, H.-B.; Zhong, H.; Karpowicz, N.; Chen, Y.; Zhang, X.-C. Terahertz spectroscopy and imaging for defense and security applications. *Proc. IEEE* **2007**, *95*, 1514–1527. [CrossRef]
7. Mansourzadeh, S.; Damyanov, D.; Vogel, T.; Wulf, F.; Kohlhass, R.B.; Globisch, B.; Hoffmann, M.; Balzer, J.C.; Saraceno, C.J. High-power lensless THz imaging of hidden objects. *IEEE Access* **2021**, *9*, 6268–6276. [CrossRef]
8. Rieh, J.-S. *Introduction to Terahertz Electronics*; Springer: Cham, Switzerland, 2021.

9. Ali, M.; Perenzoni, M.; Stoppa, D. A methodology to measure input power and effective area for characterization of direct THz detectors. *IEEE Trans. Instrum. Meas.* **2016**, *65*, 1225–1231. [CrossRef]

10. Dyakonov, M.; Shur, M.S. Shallow water analogy for a ballistic field effect transistor: New mechanism of plasma wave generation by DC current. *Phys. Rev. Lett.* **1993**, *71*, 2465. [CrossRef]

11. Knap, W.; Teppe, F.; Meziani, Y.; Dyakonova, N.; Lusakowski, J.; Boeuf, F.; Skotnicki, T.; Maude, D.; Rumyantsev, S.; Shur, M.S. Plasma wave detection of sub-terahertz and terahertz radiation by silicon field-effect transistors. *Appl. Phys. Lett.* **2004**, *85*, 675–677. [CrossRef]

12. Kojima, H.; Asano, T. Impact of subthreshold slope on sensitivity of square law detector for high frequency radio wave detection. *Jpn. J. Appl. Phys.* **2019**, *58*, SBBL05. [CrossRef]

13. Lee, H.-J.; Han, S.-T.; Yang, J.-R. CMOS plasmon detector with three different body-biasing MOSFETs. *IEEE Access* **2020**, *8*, 215840–215850. [CrossRef]

14. Kojima, H.; Kido, D.; Kanaya, H.; Ishii, H.; Maeda, T.; Ogura, M.; Asano, T. Analysis of square-law detector for high-sensitive detection of terahertz waves. *J. Appl. Phys.* **2019**, *125*, 174506. [CrossRef]

15. Sengupta, K.; Seo, D.; Yang, L.; Hajimiri, A. Silicon integrated 280 GHz imaging chipset with 4 × 4 SiGe receiver array and CMOS source. *IEEE Trans. Terahertz Sci. Technol.* **2015**, *5*, 427–437. [CrossRef]

16. Shaulov, E.; Jameson, S.; Socher, E. A zero bias J-band antenna-coupled detector in 65-nm CMOS. *IEEE Trans. Terahertz Sci. Technol.* **2021**, *11*, 62–69. [CrossRef]

17. Chai, S.; Lim, S.; Hong, S. THz detector with an antenna coupled stacked CMOS plasma-wave FET. *IEEE Microw. Wirel. Compon. Lett.* **2014**, *24*, 869–871. [CrossRef]

18. Khan, M.I.W.; Kim, S.; Park, D.-W.; Kim, H.-J.; Han, S.-K.; Lee, S.-G. Nonlinear analysis of nonresonant THz response of MOSFET and implementation of a high-responsivity cross-coupled THz detector. *IEEE Trans. Terahertz Sci. Technol.* **2018**, *8*, 108–120. [CrossRef]

19. Lee, H.-N.; Lee, H.-J.; Han, S.-T.; Yang, J.-R. Highly sensitive CMOS plasmon detector with a low-gain buffer amplifier for terahertz imaging system. *Microw. Opt. Technol. Lett.* **2021**, *63*, 2587–2591. [CrossRef]

20. Han, S.-T. Application of a compact sub-terahertz gyrotron for non-destructive inspections. *IEEE Trans. Plasma Sci.* **2020**, *48*, 3238–3245. [CrossRef]

21. Lee, M.-J.; Lee, H.-N.; Lee, H.-J.; Lee, G.-E.; Son, J.-H.; Yang, J.-R. Analysis of the effects of differential integrated antennas on voltage responsivity of sub-terahertz CMOS detectors. *IDEC J. Integr. Circuits Syst.* **2020**, *6*, 1–7.

22. Yang, J.-R.; Han, S.-T.; Baek, D. Differential CMOS sub-terahertz detector with subthreshold amplifier. *Sensors* **2017**, *17*, 2069. [CrossRef] [PubMed]

23. Karolyi, G.; Gergelzi, D.; Foldesy, P. Sub-THz sensor array with embedded signal processing in 90 nm CMOS technology. *IEEE Sens. J.* **2014**, *14*, 2432–2441. [CrossRef]

24. Lee, G.-E.; Lee, H.-J.; Han, S.-T.; Yang, J.-R. CMOS detector using customized bolt-wrench capacitor on backend oxide layer. *IEEE Microw. Wirel. Compon. Lett.* **2021**, *31*, 1012–1015. [CrossRef]

 *sensors*

Article

A CMOS RF Receiver with Improved Resilience to OFDM-Induced Second-Order Intermodulation Distortion for MedRadio Biomedical Devices and Sensors

Yongho Lee, Shinil Chang, Jungah Kim and Hyunchol Shin *

Department of Electronics Convergence Engineering, Kwangwoon University, Seoul 01897, Korea;
dldyd91@kw.ac.kr (Y.L.); starv79@kw.ac.kr (S.C.); ccog4@kw.ac.kr (J.K.)
* Correspondence: hshin@kw.ac.kr; Tel.: +82-(0)2-940-5553

Citation: Lee, Y.; Chang, S.; Kim, J.; Shin, H. A CMOS RF Receiver with Improved Resilience to OFDM-Induced Second-Order Intermodulation Distortion for MedRadio Biomedical Devices and Sensors. *Sensors* **2021**, *21*, 5303. https://doi.org/10.3390/s21165303

Academic Editors: Jong-Ryul Yang and Seong-Tae Han

Received: 30 June 2021
Accepted: 3 August 2021
Published: 5 August 2021

Publisher's Note: MDPI stays neutral with regard to jurisdictional claims in published maps and institutional affiliations.

Abstract: A MedRadio RF receiver integrated circuit for implanted and wearable biomedical devices must be resilient to the out-of-band (OOB) orthogonal frequency division modulation (OFDM) blocker. As the OFDM is widely adopted for various broadcasting and communication systems in the ultra-high frequency (UHF) band, the selectivity performance of the MedRadio RF receiver can severely deteriorate by the second-order intermodulation (IM2) distortion induced by the OOB OFDM blocker. An analytical investigation shows how the OFDM-induced IM2 distortion power can be translated to an equivalent two-tone-induced IM2 distortion power. It makes the OFDM-induced IM2 analysis and characterization process for a MedRadio RF receiver much simpler and more straightforward. A MedRadio RF receiver integrated circuit with a significantly improved resilience to the OOB IM2 distortion is designed in 65 nm complementary metal-oxide-semiconductor (CMOS). The designed RF receiver is based on low-IF architecture, comprising a low-noise amplifier, single-to-differential transconductance stage, quadrature passive mixer, trans-impedance amplifier (TIA), image-rejecting complex bandpass filter, and fractional phase-locked loop synthesizer. We describe design techniques for the IM2 calibration through the gate bias tuning at the mixer, and the dc offset calibration that overcomes the conflict with the preceding IM2 calibration through the body bias tuning at the TIA. Measured results show that the OOB carrier-to-interference ratio (CIR) performance is significantly improved by 4–11 dB through the proposed IM2 calibration. The measured maximum tolerable CIR is found to be between −40.2 and −71.2 dBc for the two-tone blocker condition and between −70 and −77 dBc for the single-tone blocker condition. The analytical and experimental results of this work will be essential to improve the selectivity performance of a MedRadio RF receiver against the OOB OFDM-blocker-induced IM2 distortion and, thus, improve the robustness of the biomedical devices in harsh wireless environments in the MedRadio and UHF bands.

Keywords: RF receiver; blocker; second-order intermodulation (IM2); orthogonal frequency division modulation (OFDM); CMOS; MedRadio; medical implanted communication service (MICS); biomedical device; biosensors

1. Introduction

An RF transceiver integrated circuit operating in the MedRadio band is widely employed for biomedical devices and sensors, as the wireless communication is essentially needed between the implanted or body-worn medical devices and outside controllers. The wireless connectivity for the biomedical devices is used to exchange the diagnostic and therapeutic data. Its non-invasiveness significantly improves the patient's comfortability by avoiding unnecessary painful surgical operations. The MedRadio band was assigned in 2009 [1], in the frequency band of 401–406 MHz providing a total 5 MHz of contiguous spectrum on a secondary and non-interference basis. The MedRadio rule was amended later in 2011 to allow networking of the devices and controllers and referred to as a medical micropower network (MMN) [2]. Since then, numerous implanted and wearable medical devices equipped with the MedRadio RF transceiver have been reported in literature, such as a wireless bio-signal monitoring system [3], an implanted cardiac defibrillator [4], an implanted pacemaker [5], an intraocular pressure monitoring system [6], physiological state monitoring prosthetic teeth [7,8], wireless position sensing system in total hip replacement surgery [9], capsule endoscopy [10,11], and so on.

The RF communication channel of the MedRadio is accessed on a shared or secondary basis. Hence, they must be robust to interferers coming from nearby other authorized primary users [12]. For mitigating the interferences, system-level techniques, such as error detection and correction via proper channel-coding, listen-before-talk (LBT), and re-transmission via a frequency monitoring and classification process, are widely employed [1]. Yet, even though the system-level interference mitigation techniques are employed, additional circuit-level techniques are still needed in the RF transceiver design to improve the overall interference resilience. Unfortunately, however, most previous MedRadio RF integrated circuits, either transceivers [4,5,12,13] or receivers [14–17], did not address those issues nor present proper circuit designs for its mitigation. Ba et al. [12] and Cha et al. [14] indeed mentioned the in-band interference issue and described the adjacent channel rejection (ACR) and the third-order intercept power (IIP3) performances. However, their studies were still limited only to the in-band interference situation and cannot be generalized to the out-of-band interference situation. It is also interesting to note that Cho et al. [18] addressed the very-high frequency (VHF) band interference issue in their dual-band RF receiver and demonstrated a blocker tolerance level of −45 dBm for the MedRadio receiver against the VHF body-channel communication interference. However, their result was also limited to the VHF-band single-tone interference issue and not applicable to the UHF-band multi-tone interference issue. It is also worth noting that non-conventional signal modulation techniques can be also effective for the interference resilience improvement. Novel approaches such as the two-tone modulation [19] and the spread-spectrum modulation [20] were proven effective in 900 MHz transceivers. Yet, they were only applicable to the on-off keying (OOK) modulation and not to the constant-envelope modulation that is more widely adopted by MedRadio devices.

The modern UHF and VHF bands are crowded with a variety of multi-tone signals. They always can be seen as unwanted strong interferers to the MedRadio biomedical devices and sensors. For example, the fifth-generation new radio (5G NR) band encompasses 600–800 MHz UHF band in its frequency range 1 (FR1) band. Additionally, digital broadcasting standards such as the advanced television systems committee (ATSC) of the United States and Korea [21], the digital video broadcasting (DVB) of Europe [22], and the intelligent service digital broadcasting (ISDB) of Japan [23] are serviced in the VHF band of 54–216 MHz and UHF band of 470–860 MHz. Moreover, these interferences are likely to adopt the orthogonal frequency division modulation (OFDM) signaling for high data rate and strong multi-path fading resilience. When an equally spaced multi-tone OFDM signal comes into the MedRadio device as an interferer, numerous intermodulation distortion components will be created by the multiple subcarrier tones, even though they are out of the band. Thus, in modern MedRadio receiver design, it is essential to analyze the effects of the out-of-band (OOB) interference on the receiver's signal-to-noise ratio (SNR) and to design circuits for mitigating its effects. Note that this issue usually cannot be neglected in typical MedRadio receivers even though the interference exists out of the band. We cannot expect sufficient filtering and attenuation for the OOB interference at the RF front end because typical MedRadio antennas show wideband characteristics [11], a complex of high-order matching and filtering at the RF front end may not be favored for low cost and small form factor realization [10], and the LNA in a receiver integrated circuit is not typically band-specific because resistive loads are frequently adopted rather than bulky inductors for the small silicon area.

In this work, we analyze the intermodulation distortion effects induced by the OOB OFDM interferer on the MedRadio receiver and, then, present a design of a CMOS RF receiver by focusing on the second-order intermodulation (IM2) distortion tolerance improvement. This work is carried out based on our prior work [17], which is expanded by adding a complex bandpass filter to realize a low-IF receiver and an IM2 calibration circuit to minimize the OOB OFDM-induced IM2 distortions.

2. Analysis of OFDM-Induced IM2 Distortion Effects

Let us assume that a desired MedRadio RF signal comes into the receiver along with an OOB OFDM blocker. Since the OFDM signal comprises multiple subcarrier tones that are independently modulated and equally spaced in the frequency domain, any combination of two subcarrier tones within the entire OFDM signal will create multiple intermodulation distortion components. Figure 1a shows the first situation such that two subcarrier tones at f_k and f_m create the third-order intermodulation (IM3) distortion at $2f_k - f_m$. The resulting IM3 falls inside the desired RF channel at f_{RF}, directly leading to SNR degradation at f_{RF}. Considering that the typical subcarrier spacing is in the range of a few 100 Hz to a few kHz for the communication and broadcasting signals, this effect is only possible when the two subcarrier tones are very close to f_{RF}. Such a close blocker signal will be likely to exist inside the desired band. Then, the in-band interference mitigation techniques such as the channel-scan-and-search-based LBT will be effective enough for its mitigation.

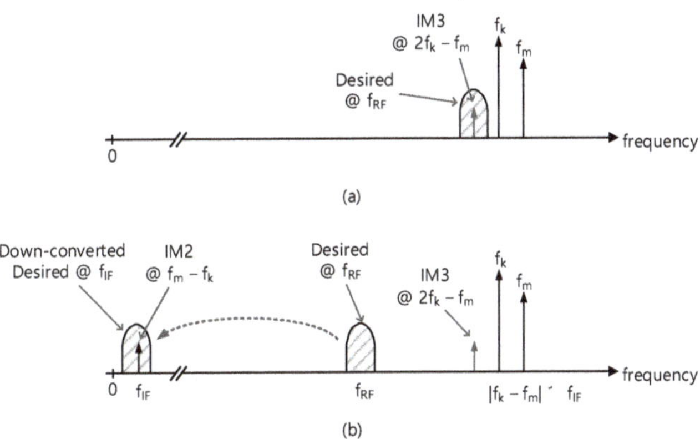

Figure 1. Two-tone blocker effects. (**a**) Third-order intermodulation effect for the close-in blocker, (**b**) second-order intermodulation effects for the far-out blocker.

On the other hand, when the OFDM blocker signal is far away from f_{RF} as shown in Figure 1b and is, thus, likely to be out of the MedRadio band, then its IM3 tone cannot directly affect the desired band at f_{RF}. Yet, the IM2 distortion at $f_m - f_k$ will fall in the down-converted IF or baseband channel at f_{IF} and lead to the SNR degradation. The LBT technique that works for the close blocker signal will not work for this situation because the blocker signal is located far away out of the band and is, thus, likely out of the channel scan range.

It should be noted that a variety of communication and broadcasting services that are being offered worldwide in the UHF and VHF bands employs the OFDM signaling. Hence, the IM2 distortions induced by the OOB OFDM blocker signal need to be analyzed rigorously to accurately assess the SNR degradation in the MedRadio receiver. Figure 2 illustrates that the OOB OFDM blocker signal at f_{BL} appears beside the desired MedRadio RF signal at f_{RF}. Let the number of subcarriers and subcarrier spacing of the OFDM blocker are N_{sub} and f_{sub}, respectively. For example, in the ATSC 3.0 standard [21], given the channel bandwidth of 6 MHz, f_{sub} is 843.7, 421.8, and 210.9 Hz depending on the OFDM FFT modes of 8K (N_{sub} = 6913), 16K (N_{sub} = 13,826), and 32K (N_{sub} = 27,649), respectively. Similarly, we can find that f_{sub} is 279–8929 Hz for DVB-T2 [22], and 992–3968 Hz for ISDB-T [23].

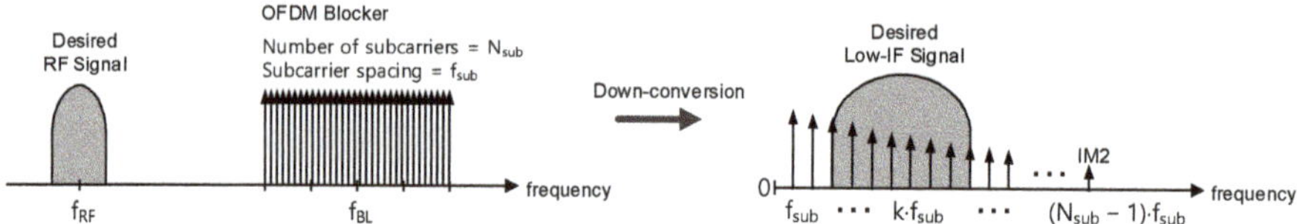

Figure 2. IM2 distortions induced by the multi-tone OFDM blocker.

As shown in Figure 2, any combination of two subcarrier tones out of the total N_{sub} subcarrier tones will create $(N_{sub} - 1)$ of IM2 tones at the baseband, starting from the first at f_{sub} to the last at $(N_{sub} - 1)\cdot f_{sub}$. If the desired IF signal band overlaps with the IM2 tones, the total IM2 distortion power can be computed by integrating its power spectral density (PSD) over the IF channel bandwidth. Then, the SNR degradation caused by the OFDM-induced IM2 distortions can be accurately evaluated. For example, the IF signal bandwidth of Figure 2 encompasses the 3rd IM2 tone at $3\cdot f_{sub}$ through the 10th IM2 tone at $10\cdot f_{sub}$. Thus, the total IM2 distortion power should be computed by integrating from the 3rd through the 10th IM2 distortion power.

The well-known input-referred second-order intercept point power (IIP_2) of an RF receiver is expressed as

$$IIP_2 = P_{in} + (P_{out} - P_{OIM2}) \tag{1}$$

where P_{in} is the input power, P_{out} is the output power, and P_{OIM2} is the output-referred IM2 distortion power, all expressed in dB or dBm. When a OFDM blocker signal that has the total power of $P_{in.total}$ is fed to a receiver having IIP_2 and a power gain of G_P, the input-referred IM2 distortion power P_{IIM2} (= $P_{OIM2} - G_P$) can be expressed by

$$P_{IIM2} = 2P_{in.total} - IIP_2 \tag{2}$$

If the total number of subcarriers is N_{sub}, the power of each subcarrier $P_{in.sub}$ is given by

$$P_{in.sub} = P_{in.total} - 10log(N_{sub}) \tag{3}$$

Now, let us compute the total input-referred IM2 distortion power by integrating the IM2 PSD over the band of interest. Ranjan et al. in their OFDM distortion analysis [24] derived an analytic expression of the IM2 PSD by assuming that IM2 tones at the baseband are uncorrelated with the original causative OFDM data. Although the exact PSD expression (Equation (12) of [24]) is not repeated here, we can find that the numerical integration of PSD can be simply carried out by collecting and adding the subcarrier PSD components at their center frequencies without a significant loss in generality and accuracy. Let us take an example of the 3rd P_{IIM2} illustrated in Figure 2, which is the first IM2 tone that exists in the IF band. We can find that the 3rd P_{IIM2} is created by $(N_{sub} - 3)$ of two subcarrier tones that are apart by $3\cdot f_{sub}$ in the original OFDM blocker signal. This observation can be generalized such that the k-th IM2 power $P_{IIM2.kth}$ can be obtained by k-th P_{IIM2} multiplied by the total number of the subcarrier pairs responsible for creating the k-th IM2 tone. It is expressed by

$$10^{\frac{P_{IIM2.kth}}{10}} = (N_{sub} - k)\cdot 10^{\frac{2P_{in.sub} - IIP2}{10}} \tag{4}$$

If the band of interest covers from the k_i-th IM2 tone through the k_f-th IM2 tone, then, the total input-referred IM2 power $P_{IIM2.total}$ that exists within the band of interest can be calculated by summing $(N_{sub} - k)$ of $P_{IIM2.kth}$ of (4) from the first k_i-th through the last k_f-th. It is expressed as follows

$$10^{\frac{P_{IIM2.total}}{10}} = \sum_{k=k_i}^{k_f}\left((N_{sub} - k)\cdot 10^{\frac{2P_{in.sub} - IIP2}{10}} \right) \tag{5}$$

Equation (5) can be written again in dB as follows

$$P_{IIM2.total} = (2P_{in.sub} - IIP_2) + 10\log\left(\sum_{k=k_i}^{k_f}(N_{sub} - k)\right) \qquad (6)$$

By substituting $P_{in.sub}$ of (6) with (3), Equation (6) can be arranged as

$$P_{IIM2.total} = 2\left(P_{in.total} - P_{offset}\right) - IIP_2 \qquad (7)$$

where P_{offset} is given by

$$P_{offset} = 10log\frac{N_{sub}}{\sqrt{\sum_{k=k_i}^{k_f}(N_{sub} - k)}} \qquad (8)$$

Equations (7) and (8) imply that the IM2 distortion power induced by the OFDM signal can be equivalently evaluated by the IM2 distortion power induced by the two-tone blocker signal as long as the two-tone blocker signal power is set to be lower than the original OFDM blocker signal power $P_{in.total}$ by P_{offset}.

This finding greatly simplifies the simulation and characterization of the OFDM-induced IM2 distortion in an RF receiver. Involvement of a multi-carrier OFDM signal in RF circuit simulations will typically require a sophisticated OFDM signal modeling, a very complex numerical analysis method, and a long simulation time. Thus, it is very time-consuming and impractical to examine the OFDM-induced IM2 effects in RF circuit design. However, if the OFDM blocker signal can be replaced simply by an equivalent two-tone blocker signal by considering an offset parameter P_{offset}, the whole simulation process will become much simpler and more convenient because we can use the conventional time- and frequency-domain circuit simulation methods. Thus, these analytic results enable us to evaluate the OFDM-induced IM2 effects very efficiently when only examining an equivalent two-tone blocker induced IM2 effects.

Let us take an example of 16K FFT mode of ATSC 3.0 [21], which has N_{sub} = 13,825 and f_{sub} = 421.875 Hz. When the IF band of a low-IF MedRadio receiver resides in 150–450 kHz, the IM2 tones appearing within the band are from k_i = 356th to k_f = 1066th components. Then, P_{offset} is computed to +6.56 dB by (8). It implies that if an ATSC blocker of −10 dBm is injected to a receiver and induces IM2 distortion at the baseband, the same amount of IM2 distortion power is also induced by a simple two-tone blocker of −16.56 dBm. Moreover, assuming the input signal power of the receiver is −50 dBm, and the receiver IIP_2 is +30 dBm, $P_{IIM2.total}$ will be −63.1 dBm (=2 × (−10 − 6.56) − 30) according to Equation (7), and the resulting SNR will be +13.1 dB as the input signal power is −50 dBm and the input-referred distortion power $P_{IIM2.total}$ is −63.1 dBm

3. Designs

Figure 3 shows the architecture of the MedRadio RF receiver. It is designed in an RF CMOS process. The receiver comprises a low-noise amplifier (LNA), transconductance (Gm) stage, quadrature down-conversion mixer with IM2 calibration, trans-impedance amplifier (TIA) with dc offset calibration (DCOC), three-stage variable gain amplifiers (VGAs), complex bandpass filter (BPF), fractional-N phase-locked loop (PLL) synthesizer, and a divide-by-4 local oscillator (LO) generator.

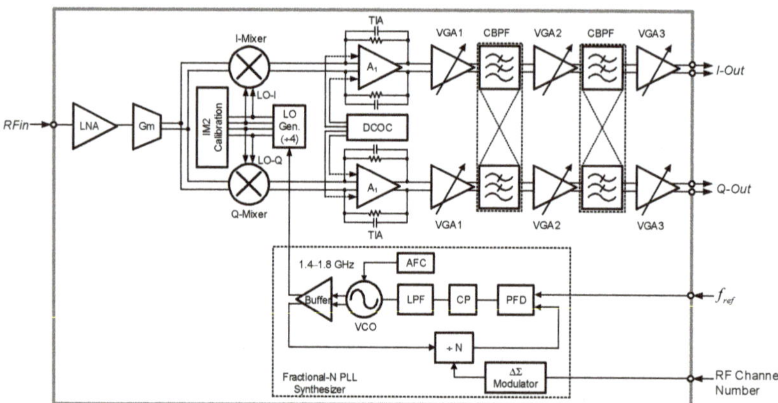

Figure 3. The MedRadio RF receiver architecture.

The receiver takes a low intermediate-frequency (IF) architecture employing a quadrature single down-conversion scheme. The IF frequency is set equal to the channel bandwidth of 300 kHz so that the down-converted IF band resides in 150–450 kHz.

The image band of the desired RF signal is rejected by the complex BPF. As shown in Figure 3, the complex BPF comprises two-stage complex biquads and three-stage VGAs. The detailed block diagram of the single-stage complex biquad is shown in Figure 4a. The unit biquad is based on a modified Tow-Thomas low-pass filter (LPF) structure in which the lossless integrator block (OPA$_1$, C$_1$) is put before the lossy integrator block (OPA$_2$, R$_3$, C$_3$). Since the second-order filtering is given by the unit biquad, the overall two-stage complex BPF presents a fourth-order bandpass filtering characteristics. Figure 4b illustrates how the low-pass characteristics of the unit biquad are translated to the complex bandpass characteristics. The cross-interconnecting resistors R$_{xa}$ and R$_{xb}$ that are placed between the I- and Q-path biquads as shown in Figure 4a shift the complex conjugate poles to real poles, resulting in the original low-pass filter response being shifted to the desired complex bandpass filter response, which gives the image rejection capability [25]. For optimal image rejection performance, the center frequency f_o is equally set to the channel bandwidth. As a result, the image component at the negative frequency band at $-f_o$ is suppressed with respect to the wanted positive frequency band at $+f_o$. To cope with process variability and also to support variable channel bandwidth modes, key performance parameters of the complex biquad are tunable by realizing the on-chip resistors and capacitors in a switched value structure. The resulting switched tuning ranges are given by 34–136 kΩ with 2-bit control for R$_1$, 82–103 kΩ with 3-bit control for R$_3$, and 0.6–7.35 pF with 4-bit control for C$_1$ and C$_3$, while R$_2$ and R$_4$ are fixed to 135 kΩ. Then, the overall complex BPF shows a tunable gain of −22–+45 dB, tunable center frequency of 0.25–3 MHz, tunable bandwidth of 0.23–2.7 MHz, and tunable quality factor of 0.9–1.1. In addition, the passband flatness is also tunable by independently controlling the on-chip R$_{xa}$ and R$_{xb}$ in the range of 78–106 kΩ so that the band-edge gain difference is adjusted in the range of −1–+1 dB [25].

Figure 4c is the schematic of the operational amplifiers OPA$_1$ and OPA$_2$. It is a fully differential two-stage amplifier with a dc common-mode feedback (CMFB). The first stage is a differential pair having *p*-FET input pair of M$_{1,2}$ and *n*-FET active load of M$_{3,4}$, and the second stage is *n*-FET common-source stage of M$_{5,6}$. Designed parameters of the FET's gate width/length are 200/0.2 μm for M$_{1,2}$, 80/0.5 μm for M$_{3,4}$, 64/0.2 μm for M$_{5,6}$, 192/0.5 μm for M$_{7,8}$, and 480/0.5 μm for M$_9$. The dc bias currents are 42 μA for M$_9$ and 18 μA for M$_{7,8}$. The frequency compensating R$_C$ and C$_C$ are 4 kΩ and 0.9 pF, respectively.

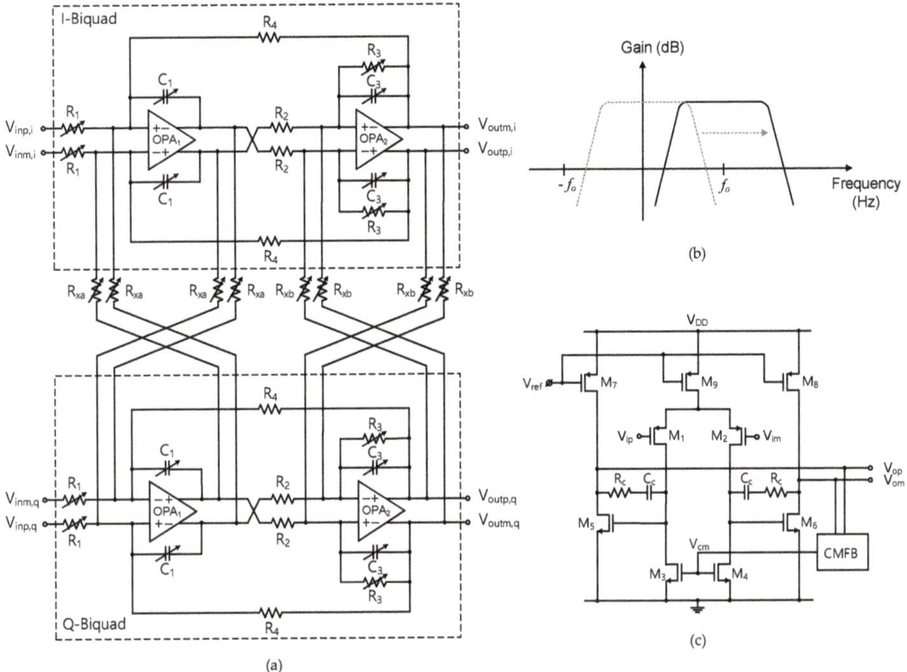

Figure 4. Complex biquad. (**a**) Block diagram, (**b**) transfer characteristics, (**c**) operational amplifier schematic.

The LO signal is generated by the fractional-N PLL frequency synthesizer. A 20-bit digital delta-sigma ($\Delta\Sigma$) modulator is employed for the fractional-N frequency generation. Details of this PLL can be found in the author's prior work [26]. The voltage controlled oscillator (VCO) covers 1.4–1.8 GHz, which is 4× higher than the MedRadio RF band of 401–406 MHz. The automatic frequency calibration (AFC) searches for an optimal sub-band out of the 32 sub-bands through 5-bit switched capacitor bank of the VCO. The divide-by-4 LO generator circuit after the VCO buffer generates I/Q LO signals with 25% duty cycle for driving the mixer FETs. The 25% duty-cycle LO signal improves the conversion gain and noise figure performances of the quadrature mixer [17].

Figure 5 shows the detailed circuit schematic of the RF front end. The LNA is based on the single-ended cascode structure with $M_{1,2}$ (gate width = 128 μm, gate length = 60 nm) and a resistive load R_D (408 Ω), dissipating 520 μA. The source degeneration inductor L_S of 9 nH and the additional gate-to-source capacitor C_{gsx} of 0.74 pF are used for simultaneous noise and power matching. The input impedance matching is achieved only by a single off-chip inductor L_{ext} (132 nH). The G_m stage with $M_{4,5}$ (gate width = 5.6 μm, gate length = 60 nm) and $M_{6,7}$ (gate width = 16 μm, gate length = 60 nm) performs a single-to-differential conversion for interfacing between the single-ended LNA and differential mixer. It also prevents a severe degradation of the LNA performance that can be otherwise caused by the low input impedance of the passive mixer. The passive mixer with $M_{11,14}$ (gate width = 14 μm, gate length = 60 nm) is adopted for the benefits of the low power dissipation and low 1/f noise. The differential I/Q LO signals $LOI_{p,m}$ and $LOQ_{p,m}$ are fed to the mixer via ac-coupling capacitors C_b (10 pF).

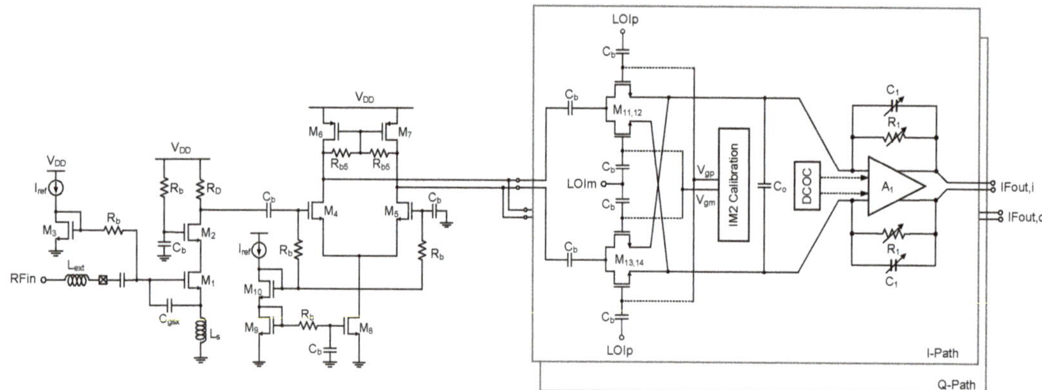

Figure 5. Circuit schematic of the RF front end.

The TIA comprises an operational amplifier A_1 and feedback components R_1 and C_1. The R_1 and C_1 are designed in a switched value structure for achieving the bandwidth tunability between 3.4 and 9.5 MHz. R_1 is switchable among 2, 8, 9, and 10 kΩ, and C_1 is switchable between 0 and 3.5 pF with 0.5 step. The A_1 is a fully differential two-stage structure with its gain bandwidth product of 21.6 MHz. The total gain of the RF front end is +42.2 dB.

The IM2 calibration at the mixer is designed to minimize the IM2 distortion. It is generally known that the IM2 distortion created by FET non-linearities is manifested by the differential mismatches in FETs, dc bias voltages, passive impedances, and layout routings. Thus, the differential mismatches need be minimized to suppress the IM2 distortion at the receiver output. In this design, as proposed in [27,28], the dc gate bias voltages V_{gp} and V_{gm} for the switch FET's M_{11-14} are controlled by employing a 6-bit R-2R digital-to-analog converter (DAC). The DAC needs to have a fine-tuning resolution for precise IM2 calibration, but a higher number of total bits will make the calibration process slow. Thus, in order to overcome the two conflicting requirements of the fine resolution and lower number of control bits, we set the full scale of the DAC to a significantly reduced value of only 50 mV, while its common-model level is tunable for a wider range between 0.5 and 0.9 V by using another 3-bit DAC. As a result, the gate bias tuning resolution is as fine as 0.78 mV. In this work, the IM2 calibration is done manually by monitoring the IM2 level with respect to the mixer gate bias voltages. In a practical mass-production stage, this IM2 calibration can be done more efficiently by using automatic test equipment (ATE).

IM2 calibration is verified through circuit simulations. Figure 6 shows simulation results. The desired RF input of -30 dBm at 402 MHz is injected, and the OOB two-tone blockers of -10 dBm at (420, 421 MHz) in one simulation and (650, 651 MHz) in the other simulation are injected. With the LO frequency set to 404 MHz, the wanted IF signal appears at 2 MHz, while the IM2 distortion appears at 1 MHz. The IM2 distortion power with respect to the wanted tone power is examined with the gate bias voltage V_{gp} tuned from 0.5 to 1.0 V, while the other gate bias V_{gm} is fixed at 0.7 V. As can be seen, the IM2 distortion level is drastically reduced when the gate bias voltage V_{gp} is set to an optimal point, which is 0.65 V for 420/421 MHz blocker and 0.7 V for 650/651 MHz blocker. It clearly implies that any unwanted mismatches that are inevitable during the circuit fabrication processes can be successfully compensated by the proposed IM2 calibration circuit, thus guaranteeing a satisfactory IM2 distortion level in the receiver.

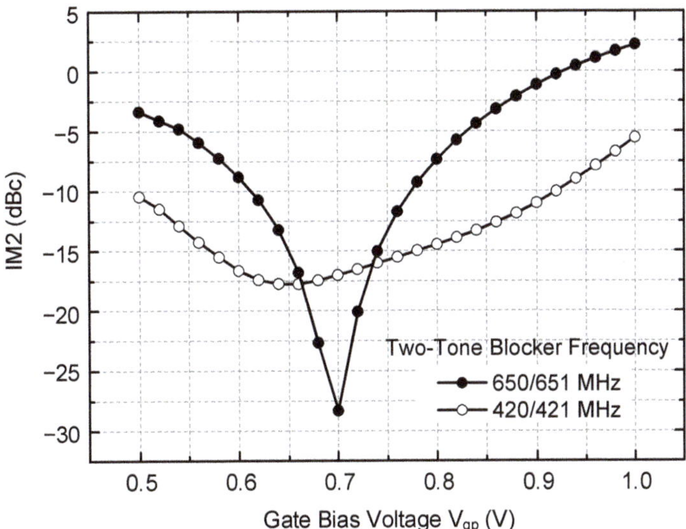

Figure 6. Simulation results of the IM2 calibration.

Meanwhile, it is found that the proposed IM2 calibration circuit creates an unwanted dc level change at the mixer output during the IM2 calibration. Since the signals from mixer output to the TIA, the first VGA, and the first biquad are all dc-coupled, the dc level changes at the mixer output will directly lead to a significant dc offset at the first biquad output of the complex BPF. The dc offset must be cancelled out even though the down-converted signal resides in the low-IF band and not in dc. It is because the residual dc offset created by the preceding IM2 calibration will harmfully reduce the dynamic range of the following stages. In the conventional low-IF receiver designs, for example, that reported in [12,29], the signals after the mixer output are ac-coupled. Then, the dc offset can be cancelled out by directly controlling the gate biases of the input FETs of a subsequent stage. However, in this design, that approach cannot be taken because the gate bias tuning will adversely alter the optimum IM2 calibration condition obtained at the previous mixer stage. Thus, in this work, we tune the body bias voltages of the input FETs of A_1 to cancel out the dc offset. This approach successfully cancels the dc offset at the biquad output, while avoiding the adverse interaction with the IM2 calibration condition at the mixer.

4. Results and Discussions

The MedRadio receiver is fabricated in a 65 nm RF CMOS process. A micrograph of the fabricated chip is shown in Figure 7, in which the major building blocks are denoted. Note that the chip includes not only the RF receiver of Figure 3 but also an RF transmitter comprising a divide-by-4 circuit and a class-D power amplifier. However, the design details and measurement results of the RF transmitter are not discussed in this paper because they are out of the scope. Nevertheless, the total die size including the entire receiver and transmitter is 2.46×1.26 mm^2. The fabricated die is mounted and directly wire-bonded on a printed circuit board for experimental measurements. A single supply voltage of 1 V is used.

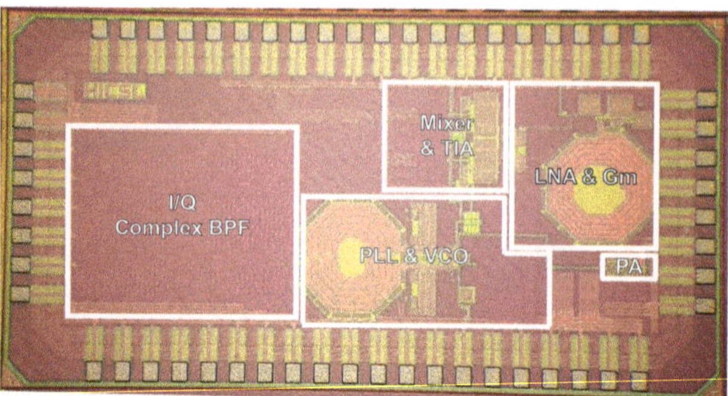

Figure 7. Chip micrograph.

Figure 8 shows the measured S_{11} of the receiver. The input impedance bandwidth having $S_{11} < -10$ dB is 37.5 MHz between 389.4 and 426.9 MHz. It shows that the single series off-chip inductor L_{ext} shown in Figure 5 is enough to achieve the satisfactory input bandwidth for the MedRadio applications. Note that this external L_{ext} is not changed during the subsequent whole OOB blocker measurements.

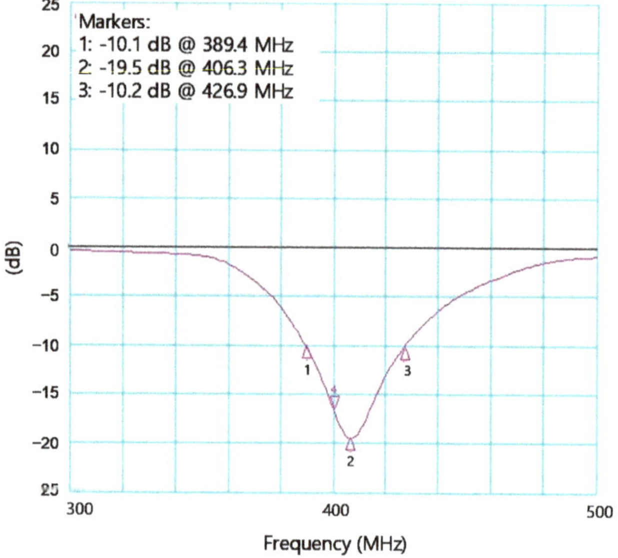

Figure 8. Measured S_{11} of the receiver.

The measured gain and noise figure performances of the RF front end are plotted in Figure 9. The measurements were done by probing intermediate test ports at the TIA output. The gain is measured by applying the RF and LO signals by using signal generators and reading the output power level by using a spectrum analyzer (N9030B of Keysight Technologies Inc., Santa Rosa, CA, USA). The measured passband gain of the RF front end is +42.2 dB. As shown earlier in Figure 5, the on-chip feedback components R_1 and C_1 of the TIA are tunable. By controlling these, the TIA bandwidth is tuned between 3.4 and 9.5 MHz in eight steps. Among them, Figure 9 only displays three selected curves corresponding to the minimum 3.4 MHz, medium 4.9 MHz, and the maximum 9.5 MHz conditions for the TIA bandwidth.

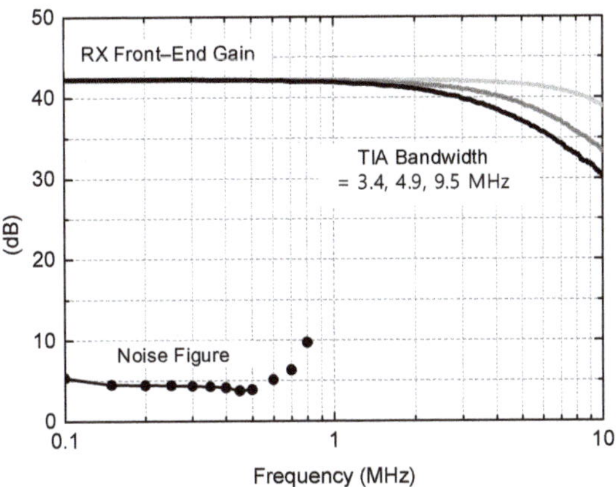

Figure 9. Measured gain and noise figure of the RF front end.

For the noise figure measurement, the total output noise power $P_{n,out}$ with the input port terminated by 50 Ω is measured by using the spectrum analyzer. With $P_{n,out}$ and the receiver's power gain G_P are known, the receiver noise figure is calculated by $(P_{n,out} + 174 \text{ dBm/Hz} - G_P)$. This method is convenient because it does not need a noise source or noise figure meter and also ensures sufficient accuracy because the noise floor level of the spectrum analyzer is more than 30 dB lower than $P_{n,out}$. The measured noise figure at the low-IF band of 150–450 kHz is given by 3.7–4.5 dB. Assuming that a non-coherent detection for binary frequency shift keying (BFSK) signal is adopted, the minimum required SNR should be 14 dB for achieving a bit error rate (BER) of 10^{-6}. In practice, the minimum BER required by a raw RF radio excluding a digital processor can be as high as $\sim 10^{-3}$ [5], and the required SNR for this can be further lower by 3 dB, that is, only 11 dB. However, in order to take various practical errors and margins into account, we decide to use 14 dB for the minimum required SNR in the following discussions. Note that the receiver sensitivity P_{sens} is given by

$$P_{sens} = -174(\text{dBm/FHz}) + NF + 10\log(B) + SNR_{min} \tag{9}$$

where NF is the receiver noise figure, B is the signal bandwidth, SNR_{min} is the minimum required SNR. With $SNR_{min} = 14$ dB, $B = 300$ kHz, and $NF = 4.5$ dB, the receiver sensitivity P_{sens} is calculated to be -100.7 dBm or 2.06 μVrms. This sensitivity performance is comparable to [12,15] and much better than [5,13].

Figure 10 shows the measured frequency responses of the complex BPF. The measured curves exhibit the bandwidth tuning and image rejection performances at a medium gain condition. Although not shown here, the gain is also tunable between -22.5 and $+45.4$ dB in 24 steps. However, normalized gains are drawn in Figure 10 for the sake of clarity. The bandwidth is tunable between 230 and 2700 kHz in 16 steps by tuning the switched resistor and capacitor components R_1, C_1, R_3, and C_3 in Figure 4. Figure 10 only displays the four selected curves out of the total 16 curves. The image rejection ratio can be evaluated by taking the ratio of the two gain values, one at the passband's center-frequency point and the other at its image-frequency point. The resulting image rejection ratio is 26.4–33.3 dB, which is sufficient enough considering that the minimum required SNR is 14 dB.

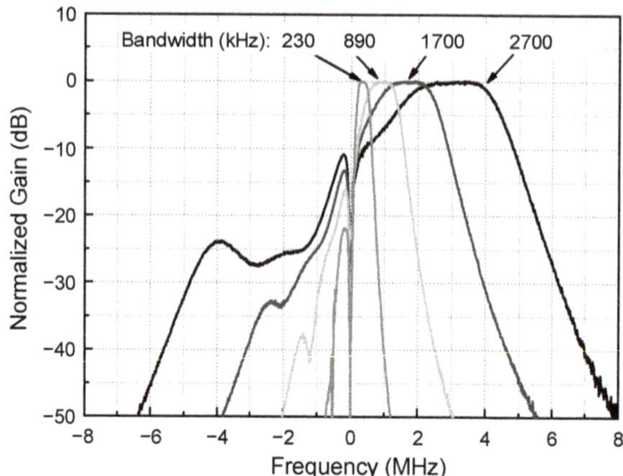

Figure 10. Measured frequency response of the complex BPF.

Figure 11 is the measured phase noise of the PLL synthesizer at the output frequency of 400 MHz. The measured phase noise is −98.7 and −125.3 dBc/Hz at 100 kHz and 1 MHz offsets, respectively. The phase noise performance is comparable to the previously reported similar LC VCOs in [12,17], whereas it is much better than [30] for the same MedRadio applications because of the LC cross-coupled structure rather than the ring oscillator structure.

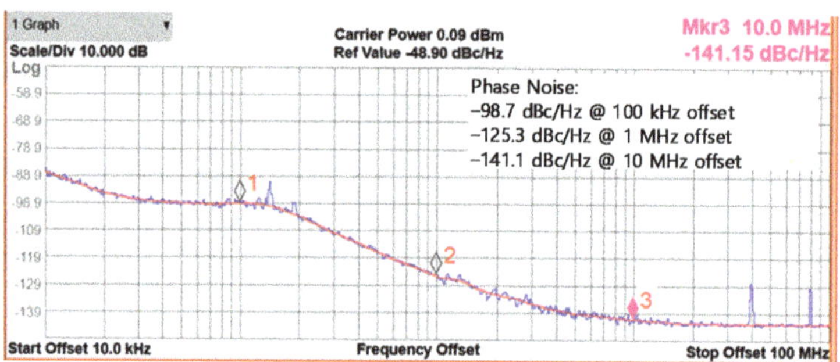

Figure 11. Measured phase noise of the PLL synthesizer.

The IM2 calibration is verified through extensive measurements over more than five samples. All results show reasonably good agreements with acceptable variability. Figure 12a shows the measured spectrum, demonstrating how the IM2 distortion is suppressed by the proposed IM2 calibration. The desired RF single tone of −90 dBm at 403.5 MHz are fed to the receiver together with two-tone blocker of −50 dBm at 433.0 and 433.2 MHz. The receiver is set to a nominal gain condition. Then, the wanted IF and unwanted IM2 tones appear at 300 kHz and 200 kHz, respectively. Before the IM2 calibration, the desired and distortion tones are −47.8 and −48.5 dBm, respectively. After the IM2 calibration is carried out, the desired tone power does not show a noticeable change, whereas the IM2 tone power is significantly reduced from −48.5 dBm to −66.1 dBm. As a result, the signal-to-IM2-distortion ratio is improved from +0.7 dB to +18.3 dB. It implies that before the calibration, the receiver cannot satisfy the 14 dB SNR requirement, but after the calibration, it successfully satisfies the SNR requirement with a sufficient margin of 4.3 dB.

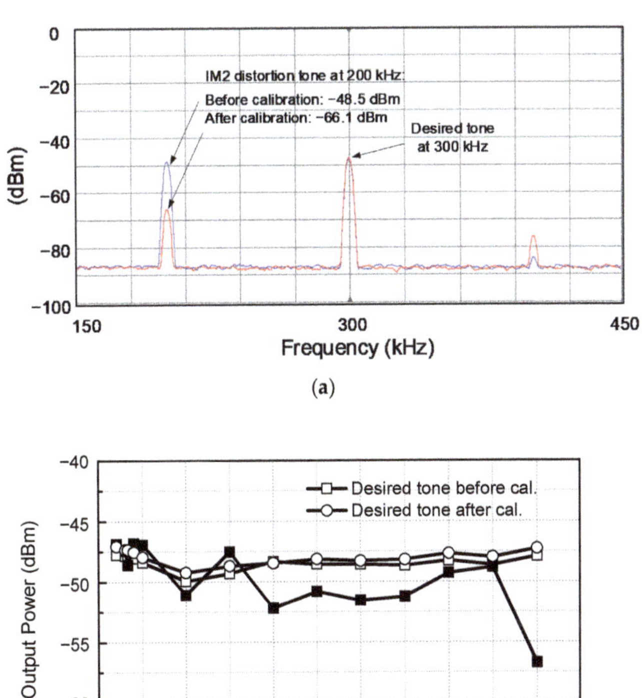

Figure 12. IM2 calibration performance measurements. (**a**) Spectrum, (**b**) results over the UHF band.

The IM2 calibration measurements are further carried out over the entire UHF band between 420 MHz and 900 MHz, and the results are plotted in Figure 12b. It is observed that the IM2 distortion is suppressed typically by 15 dB in the range of 8.3–20 dB through the IM2 calibration, while the desired tone remains almost unchanged.

The interference tolerance performances of the receiver against the OOB two-tone blocker are evaluated through the carrier-to-interference ratio (CIR) measurements and plotted in Figure 13. The maximum tolerable CIR is measured as following. First, a −90 dBm desired signal at 403.5 MHz is fed to the receiver, and an OOB two-tone blocker with 200 kHz spacing between 420 and 900 MHz is injected together. Then, the blocker signal power is raised until the output SNR reaches 17 dB. Note that we use a 3 dB higher SNR requirement for this CIR test by considering that any more noise and distortion contributions other than this IM2 distortion can be additionally involved in practice. Then, the ratio of the signal and blocker power at this 17 dB SNR condition is defined as the maximum tolerable CIR. The measured results are plotted in Figure 13. The maximum tolerable CIR indicates the receiver's selectivity performance against the two-tone blocker. As can be seen in Figure 13, the proposed IM2 calibration effectively improves the CIR by 4–11 dB across the UHF band. The CIR after the calibration is −40.2 dBc at 420 MHz, which is very close to the in-band, and gradually improves up to −71.2 dBc as the blocker frequency moves farther away out of the band. The measured CIR can also be translated to the maximum tolerable two-tone blocker power of -49.8–-18.8 dBm. Then, considering $P_{offset} = 6.56$ dB as discussed in Section 2, we can conclude that the equivalent OFDM blocker power of -43.2–-12.2 dBm can be tolerated by this RF receiver.

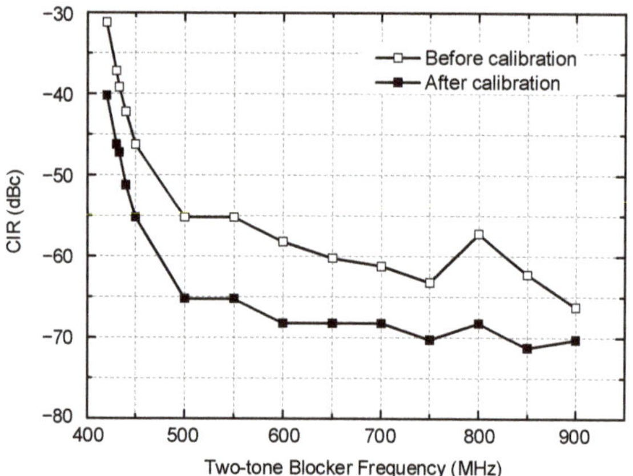

Figure 13. Measured maximum tolerable CIR against two-tone blocker.

Another important interference tolerance is against a single tone blocker. The measured results are plotted in Figure 14. The desired signal power is set to -98 dBm at 403.5 MHz, which is 3 dB higher than the sensitivity level. The single-tone blocker signal across the VHF and UHF band is injected together, and its power is raised up to a point such that the output SNR becomes 17 dB in the same reason described earlier. The CIR at a frequency that is only 4 MHz away from the desired tone is found to be -41 dBc, which corresponds to the maximum tolerable single-tone blocker power, is -57 dBm. However, in general, the maximum tolerable CIR shows much better performance as low as -70–-77 dBc for the rest of the band between 50–370 MHz and 430–900 MHz. One exception is observed at a half LO frequency near 200 MHz, where the CIR is observed to be -49 dBc. Although the MedRadio standard [1] does not clearly state this, such a limited set of exceptions would be generally accepted for the overall system operations because it can be avoided through high-level channel classification and a search process as is typically done in the Bluetooth receiver for internet-of-things (IoT) applications [28,29].

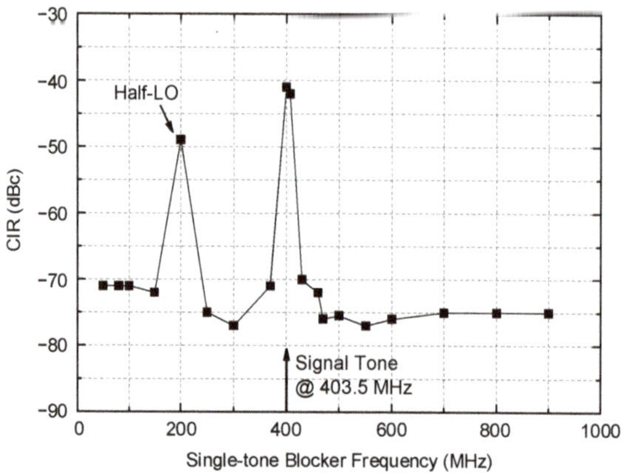

Figure 14. Measured maximum tolerable CIR against single-tone blocker.

The 1 dB gain desensitization against the single-tone blocker is measured and shown in Figure 15. The worst point appearing at 403.5 MHz corresponds exactly to the input-referred 1 dB compression power (IP_{1dB}) of the receiver, which is −40.8 dBm. As the blocker frequency moves away from the in-band, the 1 dB desensitization power continually grows from −26.7 dBm at 350 MHz to −9 dBm at 100 MHz for the lower band and also from −32.7 dBm at 450 MHz to −16.8 dBm at 900 MHz for the upper band.

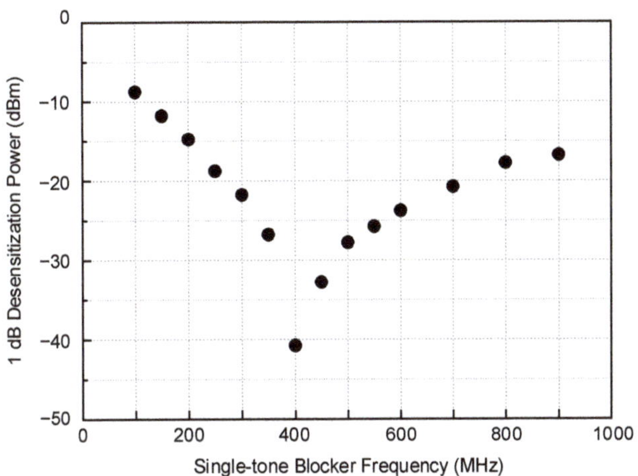

Figure 15. Measured single-tone 1 dB desensitization.

Key performances of this work are summarized and compared with previous CMOS MedRadio receivers in Table 1. It should be noted that most of the previous works [4,5,12,14–16] did not address the OOB blocker tolerance performances. Only a few works reported selectivity performances. Cho et al. [18] reported an OOB CIR of −30 dBc against the VHF band interference between 30 and 70 MHz, and Ba et al. [12] demonstrated the ACR performance of 15 dB in their MedRadio receiver. Huang et al. [19] reported the CIR of −10 dB in their 915 MHz OOK receiver. Such performances should be compared to the CIR against the single tone blocker of this work. As discussed in Figure 14, this work shows much better performance, which is better than −41 dBc without an exception, or even better than −70 dBc if two exceptions are allowed at the half LO and 4 MHz away in-band frequencies. As can be seen in Table 1, this work is the first reporting the OOB CIR performances against the two-tone and single-tone blockers, which can be translated to the maximum tolerable OOB blocker power levels. Additionally, the maximum tolerable CIR against the two-tone blocker can be translated to the maximum tolerable CIR against the OFDM blocker when the P_{offset} of Equation (8) is taken into account.

Table 1. Performance summary and comparison.

	This Work	[4]	[18]	[16]	[15]	[14]
Receiver Architecture	Low IF with complex BPF	Low IF with complex BPF	Low/Zero IF with baseband filter	Front end only	Front end only	Front end only
Gain (dB)	19.7–87.6 [1]	32	36	31	25.7	28.7
Noise figure (dB)	4.5	5.9	-	5.2	10.2	5.5
Sensitivity (dBm)	−100.7	−85	−75	-	−97	-
IIP3 (dBm)	−24	-	-	−19.5	−17	−25

Table 1. *Cont.*

	This Work	[4]	[18]	[16]	[15]	[14]
OOB CIR [2] by two-tone blocker (dBc)	-40.2–-71.2	-	-	-	-	-
OOB CIR [3] by single-tone blocker (dBc)	-70–-77	-	-30 [4]	-	-	-
Power consumption (mW)	6.78	22	8.5	0.37	1.3	0.49
CMOS process (nm)	65	180	180	180	180	180

[1] RF front end gain is +42.2 dB. [2] OOB two-tone blocker frequency at 420–900 MHz. [3] OOB single-tone blocker frequency at 50–370 MHz and 430–900 MHz with two exceptions at the half LO frequency and 4 MHz away in-band frequency. [4] OOB single-tone blocker frequency at 30–70 MHz.

5. Conclusions

The OFDM signal in the UHF band for various broadcasting and communication services can severely aggravate the selectivity performance of the MedRadio RF receiver through the IM2 distortion. We have presented an analytic investigation on how the OOB OFDM signal induces the IM2 distortion and leads to the SNR degradation at the RF receiver. We also have performed theoretical analysis on how the OFDM-induced IM2 distortion can be equivalently translated to a two-tone-induced IM2 distortion. As a result, we have introduced an offset parameter P_{offset} for compensating the difference between the multi-tone and two-tone effects. The designed MedRadio low-IF RF receiver is fabricated in a 65 nm CMOS process. Two design techniques have been described, one for the IM2 calibration through the gate bias tuning at the passive mixer's FETs, and the other for the dc offset calibration through the body bias tuning at the subsequent TIA's FETs. The proposed RF receivers have shown significant OOB CIR improvements, and the measured maximum tolerable CIR performances are between -40.2 and -71.2 dBc for the two-tone blocker and between -70 and -77 dBc for the single-tone blocker. To the author's best knowledge, this work is the first to present the analytical and experimental investigations on the OFDM-induced OOB selectivity performances in the MedRadio RF receiver for the biomedical and biosensor applications. The results of this work will be essential to enhance the selectivity performance of the MedRadio RF receiver and guarantee reliable and robust wireless communication services in implanted and wearable biomedical devices.

Author Contributions: Conceptualization, Y.L., S.C. and H.S.; circuit design and measurements, Y.L., J.K. and H.S.; formal analysis, H.S.; investigation, Y.L. and S.C.; data curation and visualization, Y.L., J.K. and H.S.; supervision, H.S. All authors participated for all other aspects. All authors have read and agreed to the published version of the manuscript.

Funding: This research was supported by the National Research Foundation of Korea under Grant No. 2020R1A2C1008484 and Kwangwoon Research Grant in 2020.

Institutional Review Board Statement: Not applicable.

Informed Consent Statement: Not applicable.

Conflicts of Interest: The authors declare no conflict of interest.

References

1. Federal Communications Commission (FCC). *Report and Order 09–23*; Federal Communications Commission (FCC): Washington, DC, USA, 2009.
2. Federal Communications Commission (FCC). *Report and Order 11–176*; Federal Communications Commission (FCC): Washington, DC, USA, 2011.
3. Srivastava, A.; Sankar, N.; Chatterjee, B.; Das, D.; Ahmad, M.; Kukkundoor, R.K.; Saraf, V.; Ananthapadmanabhan, J.; Sharma, D.K.; Baghini, M.S. Bio-WiTel: A Low-Power Integrated Wireless Telemetry System for Healthcare Applications in 401–406 MHz Band of MedRadio Spectrum. *IEEE J. Biomed. Health Inf.* **2018**, *22*, 483–494. [CrossRef]
4. Kim, J.; Choi, S.; Shin, H. A CMOS Transceiver for Implantable Cardioverter Telemetry Service. *Microw. Opt. Technol. Lett.* **2014**, *56*, 2580–2582. [CrossRef]
5. Bradley, P.D. Wireless Medical Implant Technology-Recent Advances and Future Developments. In Proceedings of the European Solid-State Circ. Conf. (ESSCIRC), Helsinki, Finland, 12–16 September 2011; pp. 37–41.

6. Ghaed, M.H.; Chen, G.; Haque, R.-U.; Wieckowski, M.; Kim, Y.; Kim, G.; Lee, Y.; Lee, I.; Fick, D.; Kim, D.; et al. Circuits for a Cubic-Millimeter Energy-Autonomous Wireless Intraocular Pressure Monitor. *IEEE Trans. Circuits Syst. I Reg. Pap.* **2013**, *60*, 3152–3162. [CrossRef]

7. Yang, C.-L.; Tsai, C.-L.; Cheng, K.-T.; Chen, S.-H. Low-Invasive Implantable Devices of Low-Power Consumption Using High-Efficiency Antennas for Cloud Health Care. *IEEE J. Emerg. Sel. Top. Circuits Syst.* **2012**, *2*, 14–23. [CrossRef]

8. Yang, C.; Zheng, G. Wireless Low-Power Integrated Basal-Body-Temperature Detection Systems Using Teeth Antennas in the MedRadio Band. *Sensors* **2015**, *15*, 29467–29477. [CrossRef]

9. Chen, H.; Gao, J.; Su, S.; Zhang, X.; Wang, Z. A Visual-Aided Wireless Monitoring System Design for Total Hip Replacement Surgery. *IEEE Trans. Biomed. Circuits Syst.* **2015**, *9*, 227–236. [CrossRef]

10. Jang, J.; Lee, J.; Lee, K.-R.; Lee, J.; Kim, M.; Lee, Y.; Bae, J.; Yoo, H.-J. A Four-Camera VGA-Resolution Capsule Endoscope System with 80-Mb/s Body Channel Communication Transceiver and Sub-Centimeter Range Capsule Localization. *IEEE J. Solid State Circuits* **2019**, *54*, 538–549. [CrossRef]

11. Kim, S.; Shin, H. An Ultra-Wideband Conformal Meandered Loop Antenna for Wireless Capsule Endoscopy. *J. Electromagn. Eng. Sci.* **2019**, *19*, 101–106. [CrossRef]

12. Ba, A.; Vidojkovic, M.; Kanda, K.; Kiyani, N.F.; Lont, M.; Huang, X.; Wang, X.; Zhou, C.; Liu, Y.; Ding, M.; et al. A 0.33 nJ/bit IEEE802.15.6/Proprietary MICS/ISM Wireless Transceiver with Scalable Data Rate for Medical Implantable Applications. *IEEE J. Biomed. Health* **2015**, *19*, 910–929. [CrossRef] [PubMed]

13. Lee, M.-C.; Karimi-Bidhendi, A.; Malekzadeh-Arasteh, O.; Wang, P.T.; Do, A.H.; Nenadic, Z.; Heydari, P. A CMOS MedRadio Transceiver with Supply-Modulated Power Saving Technique for an Implantable Brain-Machine Interface System. *IEEE J. Solid State Circuits* **2019**, *54*, 1541–1552. [CrossRef]

14. Cha, H.; Raja, M.K.; Yuan, X.; Je, M. A CMOS MedRadio Receiver RF Front-End with a Complementary Current-Reuse LNA. *IEEE Trans. Microw. Theory Techn.* **2011**, *59*, 1846–1854. [CrossRef]

15. Cruz, H.; Huang, H.; Lee, S.; Luo, C. A 1.3 mW Low-IF, Current-Reuse, and Current-Bleeding RF Front-End for the MICS Band with Sensitivity of −97 dBm. *IEEE Trans. Circuits Syst. I Reg. Pap.* **2015**, *62*, 1627–1636. [CrossRef]

16. Choi, C.; Kwon, K.; Nam, I. A 370 µW CMOS MedRadio Receiver Front-End with Inverter-Based Complementary Switching Mixer. *IEEE Microw. Wirelss Compon. Lett.* **2016**, *26*, 73–75. [CrossRef]

17. Lee, Y.; Chang, S.; Choi, J.; Moon, M.; Kim, S.; Shin, H. A CMOS Receiver Front-end with Fractional-N PLL Synthesizer for MedRadio Applications. *IEEE J. Semicond. Tech. Sci.* **2018**, *18*, 748–754. [CrossRef]

18. Cho, N.; Bae, J.; Yoo, H.-J. A 10.8 mW Body Channel Communication/MICS Dual-Band Transceiver for a Unified Body Sensor Network Controller. *IEEE J. Solid State Circuits* **2009**, *44*, 3459–3468. [CrossRef]

19. Huang, X.; Ba, A.; Harpe, P.; Dolmans, G.; de Groot, H.; Long, J.R. A 915 MHz, Ultra-Low Power 2-Tone Transceiver with Enhanced Interference Resilience. *IEEE J. Solid State Circuits* **2012**, *47*, 3197–3207. [CrossRef]

20. Ye, D.; van der Zee, R.; Nauta, B. A 915 MHz 175 mW Receiver Using Transmitted-Reference and Shifted Limiters for 50 dB In-Band Interference Tolerance. *IEEE J. Solid State Circuits* **2016**, *51*, 3114–3124. [CrossRef]

21. Advanced Television Systems Committee (ATSC). *ATSC Standard-Physical Layer Protocol*; Doc. A/322: 2018; Advanced Television Systems Committee (ATSC): Washington, DC, USA, 2018.

22. European Telecommunications Standards Institute (ETSI). *Digital Video Broadcasting (DVB)-Frame Structure Channel Coding and Modulation for a Second Generation Digital Terrestrial Television Broadcasting System (DVB-T2)*; ETSI EN 302 755 V1.4.1; European Telecommunications Standards Institute (ETSI): Sophia-Antipolis, France, 2015.

23. Association of Radio Industries and Businesses (ARIB). *Transmission System for Digital Terrestrial Television Broadcasting*; ARIB STD-B31 V2.2; Association of Radio Industries and Businesses (ARIB): Tokyo, Japan, 2014.

24. Ranjan, M.; Larson, L.E. Distortion Analysis of Ultra-Wideband OFDM Receiver Front-Ends. *IEEE Tran. Microw. Theory Tech.* **2006**, *54*, 4422–4431. [CrossRef]

25. Kim, J.; Lee, Y.; Chang, S.; Shin, H. Low-Power CMOS Complex Bandpass Filter with Passband Flatness Tunability. *Electronics* **2020**, *9*, 494. [CrossRef]

26. Lee, Y.; Kim, S.; Shin, H. A 1-V 3.8-mW Fractional-N PLL Synthesizer with 25% Duty-Cycle LO Generator in 65 nm CMOS for Bluetooth Applications. *IEEE J. Semicond. Tech. Sci.* **2018**, *18*, 730–736. [CrossRef]

27. Kaczaman, D.; Shah, M.; Alam, M.; Rachendine, M.; Cashen, D.; Han, L.; Raghavan, A. A Single-Chip 10-Band WCDMA/HSDPA 4-Band GSM/EDGE SAW-less CMOS Receiver with DigRF 3G Interface and +90 dBm IIP2. *IEEE J. Solid-State Circuits* **2009**, *44*, 718–739. [CrossRef]

28. Chang, S.; Shin, H. 2.4-GHz CMOS Bluetooth RF Receiver with Improved IM2 Distortion Tolerance. *IEEE Tran. Microw. Theory Tech.* **2020**, *68*, 4589–4598. [CrossRef]

29. Kim, S.; Kim, D.; Oh, S.; Lee, D.; Pu, Y.; Hwang, K.; Yang, Y.; Lee, K. A Fully Integrated Bluetooth Low-Energy Transceiver with Integrated Single Pole Double Throw and Power Management Unit for IoT Sensors. *Sensors* **2019**, *19*, 2420. [CrossRef] [PubMed]

30. Li, D.; Liu, D.; Kang, C.; Zou, X. A Low Power Low Phase Noise Oscillator for MICS Transceivers. *Sensors* **2017**, *17*, 140. [CrossRef] [PubMed]

sensors

Article

CMOS Implementation of ANNs Based on Analog Optimization of N-Dimensional Objective Functions

Alejandro Medina-Santiago [1,*], Carlos Arturo Hernández-Gracidas [2,*], Luis Alberto Morales-Rosales [3], Ignacio Algredo-Badillo [1], Monica Amador García [4] and Jorge Antonio Orozco Torres [5]

[1] Department of Computer Science, CONACYT-INAOE (Instituto Nacional de Astrofísica, Óptica y Electrónica), Santa María Tonanzintla, Puebla 72840, Mexico; algredobadillo@inaoep.mx
[2] Physical-Mathematical Science Department, CONACYT-BUAP, Puebla 72570, Mexico
[3] Faculty of Civil Engineering, CONACYT-Universidad Michoacana de San Nicolás de Hidalgo, Morelia 58000, Mexico; lamorales@conacyt.mx
[4] Instituto Tecnológico Superior de Rioverde, Tecnológico Nacional de México, San Luis Potosi 79610, Mexico; mony_951@hotmail.com
[5] Campus Tuxtla Gutiérrez, Tecnológico Nacional de México, Tuxtla Gutiérrez 29050, Mexico; jorge.ot@tuxtla.tecnm.mx
* Correspondence: amedina@inaoep.mx (A.M.-S.); cahernandezgr@conacyt.mx (C.A.H.-G.)

Citation: Medina-Santiago, A.; Hernández-Gracidas, C.A.; Morales-Rosales, L.A.; Algredo-Badillo, I.; Amador García, M.; Orozco Torres, J.A. CMOS Implementation of ANNs Based on Analog Optimization of N-Dimensional Objective Functions. *Sensors* **2021**, *21*, 7071. https://doi.org/10.3390/s21217071

Academic Editors: Jong-Ryul Yang and Seong-Tae Han

Received: 30 September 2021
Accepted: 19 October 2021
Published: 25 October 2021

Abstract: The design of neural network architectures is carried out using methods that optimize a particular objective function, in which a point that minimizes the function is sought. In reported works, they only focused on software simulations or commercial complementary metal-oxide-semiconductor (CMOS), neither of which guarantees the quality of the solution. In this work, we designed a hardware architecture using individual neurons as building blocks based on the optimization of n-dimensional objective functions, such as obtaining the bias and synaptic weight parameters of an artificial neural network (ANN) model using the gradient descent method. The ANN-based architecture has a 5-3-1 configuration and is implemented on a 1.2 μm technology integrated circuit, with a total power consumption of 46.08 mW, using nine neurons and 36 CMOS operational amplifiers (op-amps). We show the results obtained from the application of integrated circuits for ANNs simulated in PSpice applied to the classification of digital data, demonstrating that the optimization method successfully obtains the synaptic weights and bias values generated by the learning algorithm (Steepest-Descent), for the design of the neural architecture.

Keywords: CMOS circuit; analog system; signal processing; learning algorithm; artificial neural network

1. Introduction

The design process of complementary metal-oxide-semiconductor (CMOS) circuits consists of defining circuit inputs and outputs, hand calculations, circuit simulations, circuit layout, simulations including parasitics, reevaluation of circuit inputs and outputs, fabrication, and testing [1]. Circuit specifications are rarely defined; they can change as the design of the circuit or application progresses. This is the result of seeking the reduction of costs and improving performance of the design in its manufacture; it can also be due to the chip type or needs of the end-user. In most cases, it is not possible to make major changes to the design once the chip is in production (www.mosis.org, accessed on 18 October 2021). The characteristics of the CMOS allow the integration of logic functions with high density in integrated circuits. Due to this, CMOS has become the most widely used technology within very large-scale integration (VLSI) chips [2,3].

CMOS is used in static random access memory (RAM), digital logic circuits, microprocessors, microcontrollers, image sensors, and the conversion of computer data from one file format to another. Most configuration information on newer central processing units (CPUs) is stored on one CMOS chip. The configuration information on a CMOS chip is

113

called the real-time clock/nonvolatile RAM (RTC/NVRAM) chip, which works to retain data when the computer is shut off.

CMOS transistors are very famous because they use electrical power efficiently. They use no electrical supply whenever they are alternating from one condition to another. Furthermore, the complementary semiconductors work mutually to stop the o/p voltage. The result is a low-power dissipation VLSI design; for this reason, these transistors have changed other earlier designs like charge-coupled devices (CCDs) within camera sensors, which are used in most of the current processors. CMOS memories within a computer are a kind of non-volatile RAM that store BIOS settings, as well as time and date information.

Although most of the advances in neural networks have resulted from theoretical analysis or computer simulations, many of the potential advantages of artificial neural networks (ANNs) are expected to be implemented in hardware. Fortunately, the rapid advance in VLSI technology has made many of the previously impossible ideas now feasible to realize. Therefore, in this work, the numerical analysis of an n-dimensional objective-function optimization strategy is presented. This strategy, based on gradient descent, and its software implementation for the optimization of the weight and bias parameter values in multilayer ANNs, with the purpose of achieving their convergence, are presented as well [1,4,5]. A hardware implementation using CMOS transistors with operational amplifiers as its base elements is designed from the resulting ANN training model. A case study for the classification of digital patterns using multilayer ANNs is presented, which illustrates the implementation of the whole process. The electronic simulation of the developed integrated circuit in PSpice shows it is feasible to consider that this optimization method is efficient. In summary, the contributions of this work are as follows, sorted by relevance:

1. The design of circuits, using CMOS transistors through operational amplifiers as base elements, of the training model obtained from the parameter optimization, applied to multilayer ANNs.
2. An analysis of the designed integrated circuit (ANN hardware) based on the electronic simulation results in Pspice.
3. The software simulation of the weight and bias parameter optimization, with the purpose of achieving the convergence of a multilayer ANN (the automatic learning algorithm) based on gradient descent.
4. A multilayer ANN hardware case study simulation, designed to classify digital patterns to identify whether data is correctly sent.

Optimization techniques are ubiquitous in the field of ANNs. Generally, most learning algorithms used for training ANNs can be formulated as optimization problems because they are based on the minimization of an error function (Lyapunov function), called $E(x)$, where $x = [x_1, x_2, \ldots, x_n]^T$ is the parameter vector. One of the most important fundamentals required in the learning algorithms is the calculation of the gradient of a specific objective function $\nabla E(x)$ with respect to these parameters (synaptic weights w_{ij}, gain parameter γ_i, decay constant characteristic α_i, and Lagrange multipliers) [6–8].

It is well known that almost all optimization problems are solved numerically by iterative methods. Some of these iterative methods can be considered as a discrete-time realization of continuous-time dynamic systems. Specifically, a continuous-time (analog) dynamic system is described by a set of ordinary differential Equations (usually nonlinear), while a discrete-time dynamic system involves systems of differential equations. The advantages of using a system of differential equations are:

1. Real-world problems can be modeled with the use of differential equations, which allow humans to understand various fundamental laws of science [7]. Therefore, the simulation or implementation of a system of differential equations allows real-time optimization problems to be solved; this is due to the extraordinary parallel operation of the calculation units and the convergence properties of the neural systems.
2. The convergence properties of a continuous-time system are better because the learning rates can be set to be arbitrarily large without affecting the system stability. In con-

trast, in a discrete-time system, we must bound the control parameters in a small interval or, otherwise, the system may become unstable (i.e., the algorithm diverges).

3. A dynamic system implemented with basic differential equations exhibits more robustness to certain parameter variation [9,10].

4. Sometimes, continuous-time systems link different discrete-time iterative systems, as special cases for discretization, leading to the development of new iterative algorithms. Consequently, the simulation of continuous-time dynamic systems is more sophisticated and has faster techniques.

This work is organized as follows. Section 2 of this article shows the mathematical foundations of the Steepest-Descent objective function optimization method. Section 3 shows the analysis and development of the proposed model for the implemented circuit. Section 4 exemplify the design of integrated circuits for ANN architectures with CMOS transistors. Section 5 shows the application of the ANN to digital pattern recognition. Section 6 presents a comparison between ours and other related works. Finally, Section 7 expresses the conclusions of the work.

2. Mathematical Foundations

The basic principles of optimization were discovered in the 17th century by scientists and mathematicians such as Kepler, Fermat, Newton, and Leibniz [11,12]. Since 1950, these principles have been rediscovered for their implementation in digital computers. The progress of this effort significantly stimulated the search for new algorithms, and the field of optimization theory came to be recognized as one of the best fields of mathematics. Currently, researchers using ANNs have access to a wide range of theories that can be applied to the training of these networks.

A well-known mathematical algorithm used for solving the optimization problems is described below: the basic iterative Steepest-Descent algorithm [13,14]. The top-down method, also known as the gradient method, is defined in an n-dimensional space and is one of the oldest techniques used for minimizing a given cost function. This method forms the basis of the direct methods used in the optimization of restricted and unrestricted problems. Furthermore, it is the most used technique for non-linear optimization. In other words, what is sought is a value of x that minimizes the cost function $E(x)$.

Let us consider the following unconstrained optimization problem: find a vector $x \in \mathbb{R}^n$ that minimizes the real-valued scalar function

$$E = E(x), \tag{1}$$

This function is called the cost, objective, or energy function and x is an n-dimensional vector called the design vector. Minimizing a function is the same as maximizing the negative of the function, so there is no loss of generality in our considerations.

The point x^* is a global minimizer for $E(x)$ if $E(x^*) \leq E(x)$ for all $x \in \mathbb{R}^n$, and a strict local minimizer if the relation $E(x^*) \leq E(x)$ holds for a ball $B(x^*; \epsilon)$.

Assuming that the first and second derivates of $E(x)$ exist, a point x^* is a strict local minimizer of $E(x)$ if the gradient is zero (i.e., $\nabla E(x^*) = 0$) and the Hessian matrix is positive definite (i.e., $x^T \nabla^2 E(x^*) x > 0$).

The above statement can be formulated as a theorem on necessary and sufficient conditions for a strict local minimizer: Let $\nabla^2 E(x)$ be nonsingular for point x^*. Then, we have $E(x^*) < E(x)$ for every x in $0 < ||x - x^*|| < \epsilon$ with some $\epsilon > 0$, if $\nabla E(x^*) = 0$ and $\nabla^2 E(x^*)$ is symmetric and positive definite.

Considering an optimization problem, which has a cost function:

$$E(x) \text{ subject to } x \in \mathbb{R}^n, \tag{2}$$

Suppose there exists, at the same time, a gradient vector and the Hessian matrix of the objective function (i.e., they can be evaluated analytically), then this involves generating a sequence of search points $x^{(k)}$ through the iterative procedure:

$$x^{(k+1)} = x^{(k)} + \eta^{(k)} * d_k, \; x(0) = x^{(0)}, \; (k = 0, 1, 2, \ldots), \tag{3}$$

where $\eta^{(k)} > 0$ determines the length of the step (learning rate) to be taken in the direction of the vector $d_k \cong \Delta x^{(k)}$ (search direction). In numerical optimization, there are different techniques to calculate the parameter $\eta^{(k)}$ and the direction of the vector d_k. For convenience, four basic methods are presented:

1. Gradient method (Steepest-Descent) [6,12,15], where the direction is defined as:

$$d_k := -\nabla E(x^{(k)}), \tag{4}$$

2. Newton's method [11], in which the search direction is determined by:

$$d_k := -[\nabla^2 E(x^{(k)})]^{-1} \nabla E(x^{(k)}), \tag{5}$$

3. Since the calculation of the Hessian inverse matrix $[\nabla^2 * E(x^{(k)})]^{-1}$ can be slightly complicated, a symmetric positive matrix of dimension $n \times n$ is defined, which is called H_k.

$$H_k \cong [\nabla^2 E(x^{(k)})]^{-1}, \tag{6}$$

Applying the quasi-Newton method [11,12,15] the search direction is determined by:

$$d_k = -H_k \nabla E(x^{(k)}), \tag{7}$$

4. The conjugate gradient method [11,12,15–17] calculates the current search direction d_k as a linear combination of the current gradient vector and the previous search direction. This simple way to find the direction is calculated by:

$$d_k = -\nabla E(x^{(k)}) + \beta_k d_{k+1}, \; (k = 0, 1, 2, \ldots), \tag{8}$$

with $d_0 = -\nabla E(x^{(0)})$, where β_k is a scalar parameter that ensures that the vector sequence d_k satisfies the condition of mutual conjugation.

The length of the step $\eta^{(k)}$ is usually determined by these methods using one of the following techniques: (1) Minimization along the line, and (2) A fixed step size in one dimension. In the next subsection, the mathematical development of the continuous-time interactive algorithm for calculating the local minimum of a cost function is described.

2.1. Continuous-Time Iterative Algorithm

The gradient descent method discussed above can be written as:

$$x^{(k+1)} := x^{(k)} - M_k \nabla E(x^k), \tag{9}$$

where M_k is symmetrically defined as a positive matrix of dimension $n \times n$. The appropriate choice of matrix M_k is critical, based on the convergence properties of the algorithm. The discrete-time minimization algorithm determines the local minimum of the cost function $E(x)$ as the limit of the sequence $E(x^{(k)})(k = 0, 1, 2, \ldots)$ where $x^{(0)}$ is an initial estimate of the local minimizer X^*.

These iterative algorithms generate a sequence of points $\{x^{(k)}\}$ and a search direction $d_k \cong \Delta x^{(k)}$ through a discrete-time approximation for some continuous-time trajectory from the starting point $x^{(0)}$ to the minimum point x^* (stationary). The continuous-time trajectory $x(t)$ is usually determined by a system of differential equations as follows:

$$\frac{dx_j}{dt} = -\sum_{i=1}^{n} \mu_{ji} \frac{\partial E(x)}{\partial x_i}, \; X_j(0) = x_j^{(0)}, \; (j = 1, 2, \ldots, n), \tag{10}$$

This expression can be written more compactly in the form of a matrix:

$$\frac{dx}{dt} = -\mu \nabla E(x), \; x(0) = x^{(0)}, \tag{11}$$

where $x(t) \in \mathbb{R}^n$, $\mu(x, t)$ is a positive definite matrix of dimension $n \times n$ whose inputs are generally dependent on time, and the variable $x(t) = [x_1(t), x_2(t), \ldots, x_n(t)]^T$. Determining the positive matrix μ requires system stability. The next subsection describes a basic gradient system used to determine the direction of gradient change of a system of differential equations.

2.2. Basic Gradient System

Considering the simple case for which the matrix $\mu(x, t)$ is reduced to a positive scalar function $\mu(t)$, the system of differential equations according to (10) is simplified to:

$$\frac{dx_j}{dt} = -\mu(t) \frac{\partial E}{\partial x_j}, \tag{12}$$

with $x_j(0) = x_j^{(0)}$ $(j = 1, 2, \ldots, n)$; $\mu(t) > 0 \in C$ is a constant and theoretically it can be a large arbitrary set; the learning relationship $\eta^{(k)}$, in a discrete-time slope algorithm (ascending slope), is limited to a small interval to ensure that the algorithm converges. The previous system of differential equations is called the basic dynamic gradient system. Continuous-time gradient downward methods employ (12), since the search of $-\nabla E(x)$ is in the direction of maximum negative change of the objective function $E(x)$ at some point.

An interesting aspect of the method is the fact that the direction determined by the discrete-time iterative algorithm is slightly oscillatory. In contrast, the direction obtained by the continuous-time gradient method is monotonous. In other words, the oscillation effects can be eliminated using a continuous-time gradient system. As described, the main objective of this algorithm is to find a value of x that minimizes $E(x)$.

3. Analysis and Development

From the optimized model of the ANN, we carried out the CMOS circuit design based on the operational amplifier to design the base cell called perceptron, which is commonly used in multi-layer perceptron (MLP) architectures. Figure 1 shows the block diagram of the proposed methodology for the development of the circuit. Block 1 represents the numerical analysis stage and the simulation of the optimization of the objective function (obtaining synaptic weights and bias) for the ANN, based on gradient. Block 2 is focused on the design of the base neuron (perceptron) at the circuit level. Finally, block 3 shows the development and simulation of the complete MLP circuit to obtain the neural network behavior proposed in the case study.

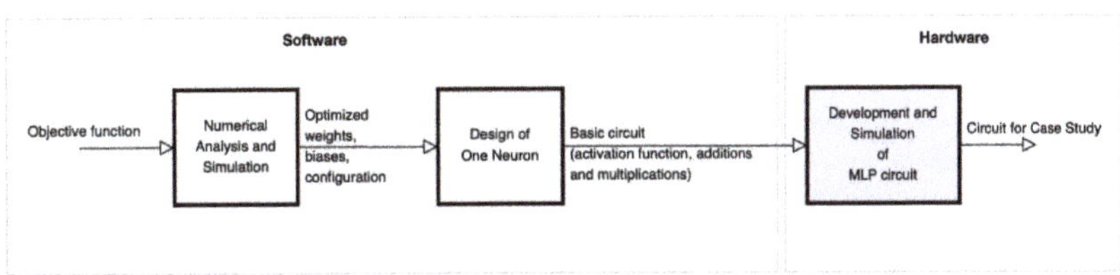

Figure 1. Block diagram of the design and development of the proposed methodology.

All the optimization algorithms described above employ a system of first-order differential equations. Now, we need to apply the optimization to a system of second-order differential equations. Therefore, to improve the convergence properties, we can use a system of higher-order ordinary differential equations by considering the system of second-order differential Equations (1), (3) and (8) as follows:

$$\delta(t)\frac{d^2x}{dt^2} = -\gamma(t)T\frac{dx}{dt} - \nabla E(x), \tag{13}$$

with initial conditions $x(0) = x^{(0)}$, $(\frac{dx}{dt})(0) = x'(0) = x'^{(0)}$, where $x(t) \in \mathbb{R}^n$, $\delta(t)$ and $\gamma(t)$ are positive functions with real values for $t \geq 0$, and $T = [T_{ij}]$ is a positive definite symmetric matrix of dimension $n \times n$.

A simple case is when δ, γ, and T are constant; for example: $\delta(t) = \delta_0 \geq 0$, $\gamma(t) = \gamma_0 > 0$, $T = [T_{ij}]$, and T_{ij} is constant for $i, j = 1, 2, \ldots, n$. However, in general, the matrix $T = T(x, t)$ depends on time. A particular choice of the parameters of (13) makes it possible to obtain almost all first-order methods. For example, when $\delta(t) = 0$, the conditions are the following:

1. For $T = I$ (identity matrix) and $\gamma(t) = \mu^{(-1)} > 0$, the gradient descent method is applied.
2. For $T = \nabla^2 E(x)$ (Hessian matrix) and $\gamma(t) = 1$, Newton's method is applied.
3. For $T = [\nabla^2 E(x) + v(t)I]$ and $\gamma(t) = 1$, the Levenberg–Marquardt method [11,12,15–17] is applied.

The system of second-order equations is inspired by classical mechanics and has the following physical interpretation [6]: (13) represents Newton's second law (mass × acceleration = force) for a mass particle $\delta(t)$ moving in a space $\mathbb{R}^n$ subject to a force $-\nabla E(x)$ given by the potential $E(x)$ and with force $-\gamma(t)T(\frac{dx}{dt})$. Since $\gamma(t) > 0$, the force $-\gamma(t)T(\frac{dx}{dt})$ is dissipative and $\gamma(t)$ is the coefficient of friction. Generally, the mass coefficient of $\delta(t)$ and the coefficient of friction $\gamma(t)$ are constant in time or tend to zero as time approaches infinity.

The application of a system of second-order differential equations has several important advantages over a system of first-order differential equations, which are:

1. Because of the initial force, the local minimum of the objective function $E(x)$ can be avoided by an appropriate choice of parameters, and the network can find an overall minimum, although this cannot be guaranteed.
2. The second-order differential equations have better flexibility. For example, for the same starting point $x(0)$ different from the selection of $(\frac{dx}{dt})(0)$, they can lead to a different local minimum. That is, we changed coefficients γ and δ making it possible to reach a local minimum from the same initial condition $(x(0), x'(0))$. Thus, in a system of differential equations of the form given by (13), an additional control of the solution is provided.
3. A system of second-order differential equations may have a better property of convergence. Therefore, responses can be obtained in its trajectory.

To evaluate the behavior of the gradient descent algorithm, we performed two simulations. In the first example, we have a single-variable objective function for which the determination of the local minimum depends on the initial condition of the variable. A two-variable objective function (which can fall into a saddle point or a local minimum) is described in the second example. This case can be avoided by an appropriate choice of the initial condition of the variables. The results obtained from the applications of the algorithm and the initial conditions are shown in Tables 1 and 2, respectively.

Table 1. Part One: Objective function.

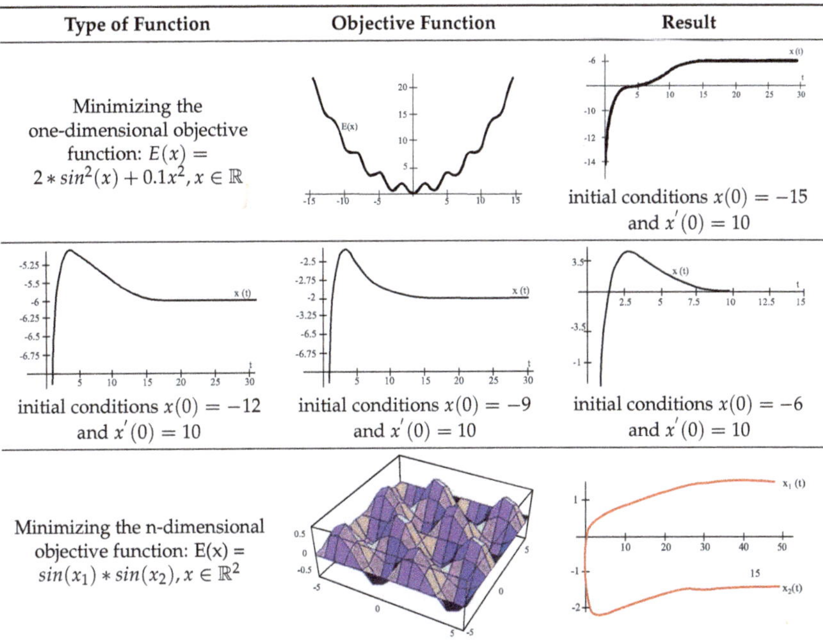

Type of Function	Objective Function	Result
Minimizing the one-dimensional objective function: $E(x) = 2 * sin^2(x) + 0.1x^2, x \in \mathbb{R}$		initial conditions $x(0) = -15$ and $x'(0) = 10$
initial conditions $x(0) = -12$ and $x'(0) = 10$	initial conditions $x(0) = -9$ and $x'(0) = 10$	initial conditions $x(0) = -6$ and $x'(0) = 10$
Minimizing the n-dimensional objective function: $E(x) = sin(x_1) * sin(x_2), x \in \mathbb{R}^2$		

Applying (13), the optimization problem (one-dimensional) is:

$$\delta_0 \frac{d^2x}{dt^2} = -\gamma_0 \frac{dx}{dt} - \mu_0(4 * sin(x)cos(x) + 0.2x), \tag{14}$$

Applying (13), the optimization problem (n-dimensional) has the following form:

$$\delta_0\left(\frac{d^2x_1}{dt^2}\right) = -\gamma_0\left(\frac{dx_1}{dt}\right) - \mu_0 * cos(x_1) * sin(x_2), \tag{15}$$

$$\delta_0\left(\frac{d^2x_2}{dt^2}\right) = -\gamma_0\left(\frac{dx_2}{dt}\right) - \mu_0 * cos(x_2) * sin(x_1), \tag{16}$$

Table 2. Part Two: Initial Conditions of the Objective functions.

Type of Function	Initial Conditions of Objective Function
Minimizing the one-dimensional objective function	where μ_0 is a scaling factor. The coefficients δ, γ, and $T(=1)$ are constant. For $\delta_0 = 0.1$, and $\gamma_0 = 0.15$, the initial conditions (starting points) are: $x(0) = -15$, -12, -9, -6, and $x'(0) = 10$.
Minimizing the n-dimensional objective function	with initial conditions $x_1(0) = -1$, $x'_1(0) = 2$, $x_2(0) = -1$, $x'_2(0) = -2$.

When a search of the local minimum of a particular objective function of two or more variables is performed, it is necessary to avoid falling into the saddle points caused by the initial conditions of the variables when calculating the solution. In Table 1, the results obtained for examples 1 and 2 show how we can improve the convergence properties of an objective function with one or two variables. In general, an n-dimensional objective function will work, using the proposed system of higher-order ordinary differential equations. In the following section, we will show how this approach can be applied. We will do this through a case study for its application to an ANN implemented using analog systems and CMOS circuits.

4. Circuit Design: A Case Study of ANNs Implemented in CMOS Circuits

In this section, the application of the proposed system to an ANN [1,4,5] is described. An analog multiplier multiplies the outputs of a voltage adder concerning a constant. The voltage adder generates the sum of the voltages of the matrix W_{ij}, which represents the synaptic weight matrix (Figure 2), and the signal of the non-linear function generator. We used operational amplifiers (op-amps) to design each base neuron that composes the neural architecture. It means that we implemented the structure of the neural circuit of Figure 2 to produce the ANN represented in Figure 3. We remark that each neuron node has two parts: a sum function and an activation function (sigmoid). We implemented the first one with an op-amp inverting adder (designed based on Figure 4) and the second one with an array of op-amps and voltage limiters (diodes).

Figure 2 shows the block diagram for the function, which consists of two continuous-time integrators (whose response depends on the feedback network), one analog multiplier, one summing amplifier, and one non-linear function generator for calculating the gradient of the objective function at the circuit level. The optimized parameters $x(t) = [x_1(t), x_2(t), \ldots, x_n(t)]^T$ are the output signals of the integrator. This circuit is characterized by having a more robust output (insensitive to small perturbations) with respect to the parameter variation. The function generator is the only one that precisely calculates the gradient of the objective function.

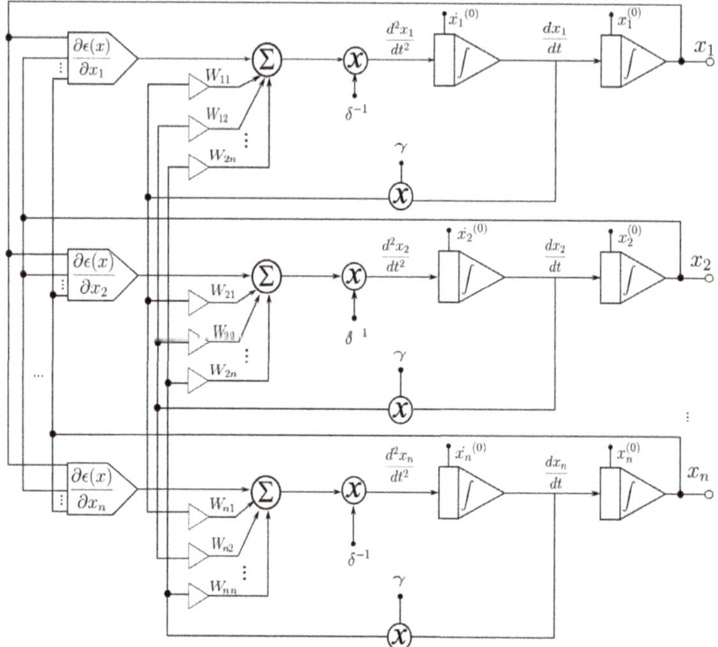

Figure 2. Block diagram for applications to ANNs.

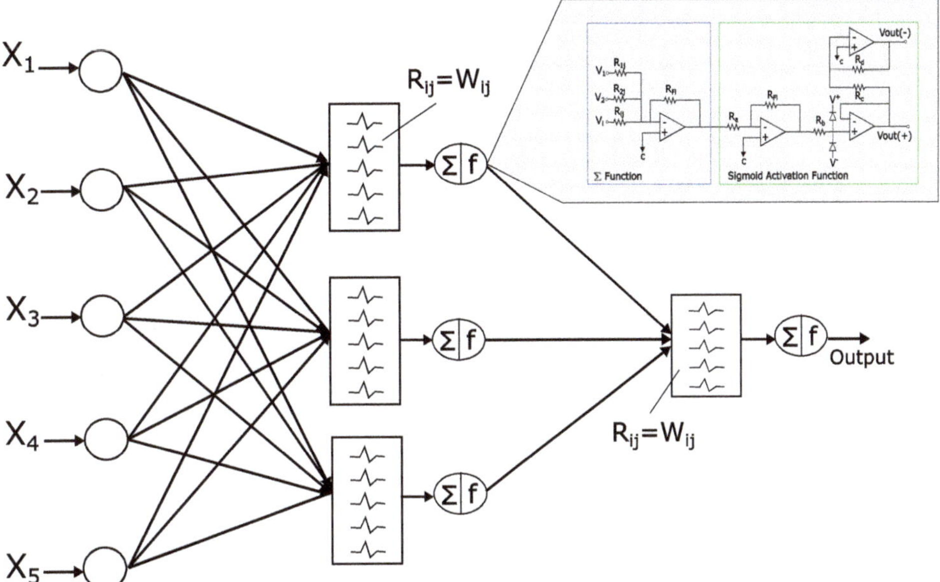

Figure 3. Complete circuit based on ANN.

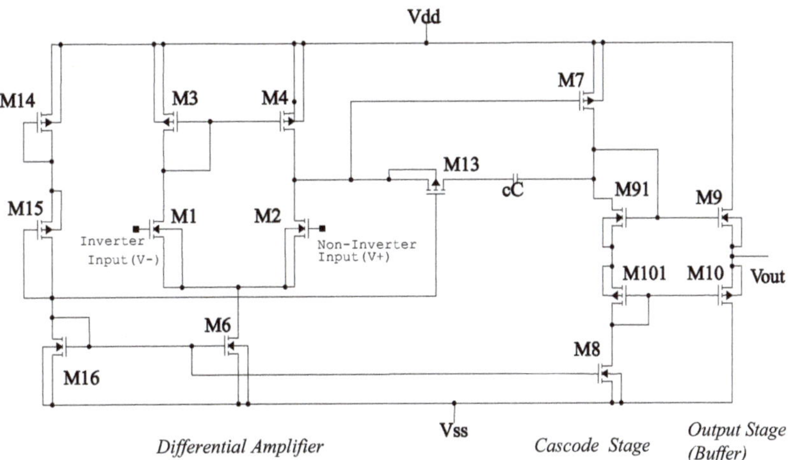

Figure 4. Operational amplifier 1.2 μm.

According to Figure 2, the design of the analog neural network consists of the development of the integrator circuit. As shown in Figure 5, the integrator circuit uses the op-amp shown in Figure 6 (the block diagram of the 1.2 μm technology operational amplifier in Figure 4). The op-amp is designed according to the specifications described in Table 3.

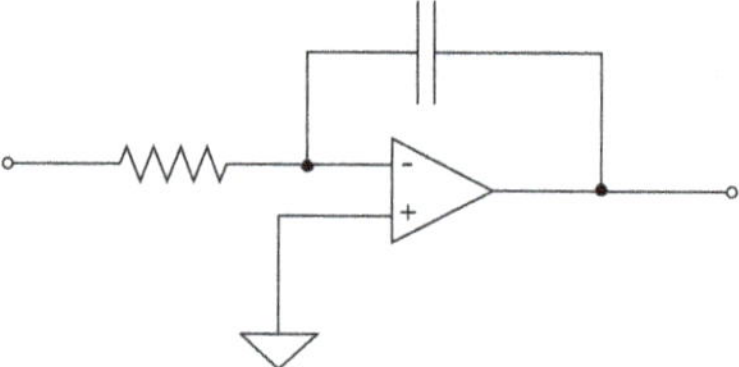

Figure 5. Integrator Circuit.

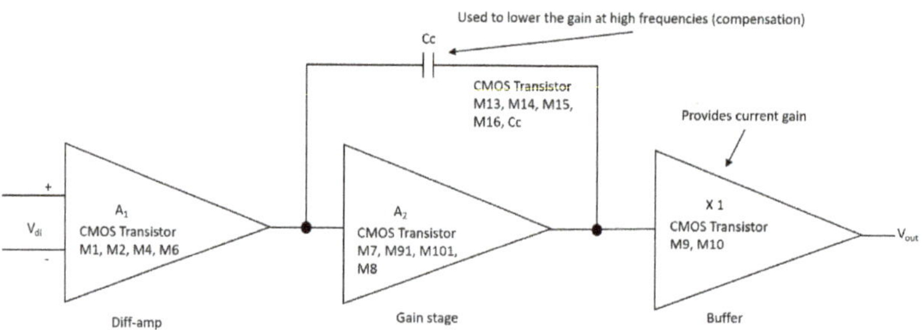

Figure 6. Block diagram of the 1.2 μm operational amplifier.

Table 3. Parameters of CMOS op-amp.

Parameter	Value
Open-loop voltage gain	60 dB
Phase margin (ϕ_o)	81°
Unity gain bandwidth (GB)	2.30 MHz
Slew rate (SR)	21.06 V/μs
CMRR	47 dB
Offset voltage (Vos)	284 μV
Compensation capacitance (CC)	8.75 pF
Capacitive load (CL)	20 pF
Power supply (VDD, VSS)	+/−2.5 V
Power consumption	1.28 mW
Common mode input voltage range	−608.696 mV to 1.98 V

Figure 6 is, in detail, the block diagram of the 1.2 μm technology operational amplifier in Figure 4 that shows the operational amplifier with CMOS transistors, formed by the stages of a differential amplifier (M1, M2, M4, and M6), gain stage (M7, M91, M101, and M8) and the output stage (buffer) (M9 and M10). Compensation network M13 and capacitor Cc, where M13 is made up of a polarization network by the transistors M14, M15, and M16. The dimensions of the main semiconductor N-channel and P-channel transistors are listed in Table 4. The characteristic values of the integrator are described in Table 5 [18,19].

The complete circuit of the analog neural network is shown in Figure 3, where the inputs X_i, the weights (defined by the resistors $R_{ij} = w_{ij}$), and the activation functions (Σ and F) are represented. The basic circuits that correspond to the analog neural network neurons are the inverting amplifiers and the activation function. The inverting amplifier circuit is used as a summation block. Hence, when many input voltages are connected to the inverting input terminal, the resulting output is the sum of all the input voltages applied, although inverted; this output, combined with the feedback resistor, generates the multiplication by a weight. The circuit for implementing a neuron is shown in Figure 7, where the function Σ computes multiplications and the activation function is a *Sigmoid* [19].

Table 4. Widths and lengths for N-channel and P-channel MOS transistors.

Transistor	2 µm Technology N-Channel MOS		1.2 µm Technology P-Channel MOS	
	Width	Length	Width	Length
Differential Stage				
M1	15 µm	5 µm	9 µm	3 µm
M2	15 µm	5 µm	9 µm	3 µm
M3	70 µm	5 µm	42 µm	3 µm
M4	70 µm	5 µm	42 µm	3 µm
M6	30 µm	5 µm	18 µm	3 µm
M16	15 µm	5 µm	9 µm	3 µm
M14	70 µm	5 µm	42 µm	3 µm
M15	70 µm	5 µm	42 µm	3 µm
Cascode Stage				
M13	70 µm	5 µm	42 µm	3 µm
M7	70 µm	5 µm	42 µm	3 µm
M91	15 µm	2 µm	9 µm	1.2 µm
M101	70 µm	2 µm	42 µm	1.2 µm
M8	15 µm	5 µm	9 µm	3 µm
Output Stage				
M9	150 µm	2 µm	90 µm	1.2 µm
M10	700 µm	2 µm	420 µm	1.2 µm

Table 5. Characteristic Values of the CMOS integrator circuit.

Characteristic	Values
Slew rate	21.06 V/µs
Cutoff frequency	2.2570 MHz
Gain	60 dB
Phase Margin	83.782°
A1 gain	33.33
A2 gain	55.55

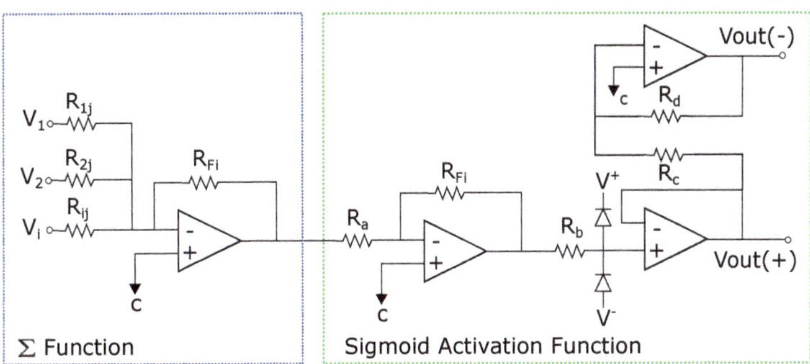

Figure 7. Circuit for one neuron.

As shown in Figure 3, the architecture has 9 neurons (configuration 5-3-1), and each neuron has 4 op-amps due to the adder and the sigmoid activation function (see Table 6). The total power consumption is 46.08 mW, because there are 36 op-amps for the entire circuit, and each of them consumes 1.28 mW [18,20,21].

Table 6. Number of op-amps and total power consumption.

Description	Quantity
No. of op amps per adder	1
No. of op amps per activation function	3
No. of op amps per neuron	4
No. of neurons for the 5-3-1 architecture	9
No. of op amps for the 5-3-1 ANN circuit	36
Total power consumption	46.08 mW

5. Application to an Implemented Analog ANN for Pattern Detection

The ANN-based architecture of this work (see Figure 3) is proposed for pattern recognition as an application example, where the recognition task consists of detecting a value of +1 or −1 for Hamming code correction. One of the advantages of the proposed circuit is that it can be integrated as a module in some communication systems in application-specific integrated circuits (ASIC).

The electronic simulation of the backpropagation neural network is implemented on the basic circuits designed for a typical neuron (the weighted sum function and the activation function). This simulation is carried out using the Pspice program [18], which allows changing the device parameters as well as the stimuli, in such a way that different operating conditions of the circuit are covered.

Once modeled the behavior of the cells that make up the backpropagation, the complete circuit simulation to study the network performance as a whole can continue. This is done by connecting the synapses and the activation functions according to the configuration established in Figure 3. The input vector is bipolar (see Table 7).

Figure 8 represents the neural network response (designed to classify input patterns with a supervised learning algorithm) to different stimuli, with a sweep of the output signal concerning the input signal. Figure 8a represents a stimulus P = [−1 1 1 1 1] and the corresponding output T = [1]; if we observe from left to right, we have a negative input and a positive posterior one, which is consistent with the input pattern and its positive output: 1. Figure 8b shows the behavior of the architecture with a stimulus P = [−1 −1 −1 −1 −1] and an output T = [1]; if we sweep the input signal from left to right, we can observe that all the input patterns are negative and we have a positive output at 1. For Figure 8c we have an input pattern P = [1 −1 −1 −1 −1] and an output T = [−1], sweeping from left to right the input pattern is negative and then positive; where the first position of the input vector will be the least significant one for the response of the architecture, resulting in a negative output. Finally, for Figure 8d we have an input pattern P = [1 −1 1 1 1] and an output T = [−1]; if we sweep the output signal from left to right, we observe that the negative value of the pattern is not preponderant for the output of the architecture, which in this case is negative.

The proposed methodology is based on the problem analysis, the datasets definition (inputs and outputs) for a supervised method, the evaluation of ANN configurations, the selection of the best configuration, the implementation of the configuration through operational amplifiers, and the simulation of the ANN-based circuit. In this work, a detection circuit is implemented. The outline training is done through the use of Mathematica (i.e., the training, in which weights and biases are estimated, is a software-based step carried out prior to the execution of the simulation of the ANN-based circuit), and the ANN configuration is carried out using op-amps.

Table 7. Bipolar input vector.

Input Vector (P)					Target (T)
−1	−1	−1	−1	−1	1
−1	−1	−1	−1	1	1
−1	−1	−1	1	−1	1
−1	−1	−1	1	1	1
−1	−1	1	−1	−1	1
−1	−1	1	−1	1	1
−1	−1	1	1	−1	1
−1	−1	1	1	1	1
−1	1	−1	−1	−1	1
−1	1	−1	−1	1	1
−1	1	−1	1	−1	1
−1	1	−1	1	1	1
−1	1	1	−1	−1	1
−1	1	1	−1	1	1
−1	1	1	1	−1	1
−1	1	1	1	1	1
1	−1	−1	−1	−1	−1
1	−1	−1	−1	1	−1
1	−1	−1	1	−1	−1
1	−1	−1	1	1	−1
1	−1	1	−1	−1	−1
1	−1	1	−1	1	−1
1	−1	1	1	−1	−1
1	−1	1	1	1	−1
1	1	−1	−1	−1	−1
1	1	−1	−1	1	−1
1	1	−1	1	−1	−1
1	1	−1	1	1	−1
1	1	1	−1	−1	−1
1	1	1	−1	1	−1
1	1	1	1	−1	−1
1	1	1	1	1	−1

In [22,23], the authors presented a circuit-level implementation of the backpropagation learning algorithm, exploring the gradient descent method in the design of analog circuits. They used an ANN in memristive crossbar arrays, simulating in SPICE on TSMC's 180 nm CMOS technology, reporting output voltages and timing behaviors.

In our case, we presente an analog system simulation based on the solution to optimization problems with objective functions of one and two variables using the gradient descent method (ascending slope). We showed the ANN configuration behavior (learning algorithm), where its input and process signals are analog. Our circuit-level implementation was made with a 5-3-1 configuration using the backpropagation training algorithm, reporting the simulation-level architecture results in PSpice level 9.1 on 1.2 µm technology with a lambda of 0.5 µm, with widths and lengths of the X-channel defined by the technology.

Although the learning of the neural network was developed offline, and we consider the optimization results obtained in the mathematical part of the article were satisfactory, we emphasize that its essential purpose was to illustrate the applicability to optimize the learning algorithm of an ANN. We also simulated the optimized ANN model through the design of a circuit with CMOS transistors.

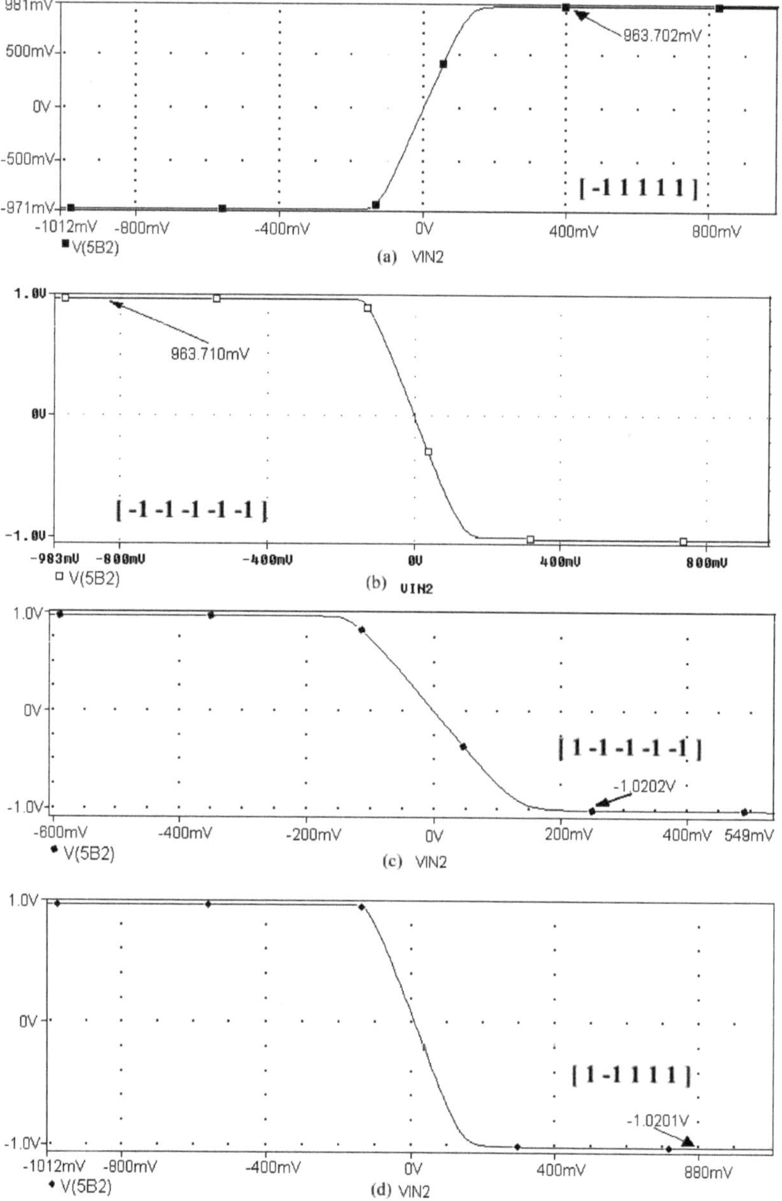

Figure 8. Response of the neural network to different stimulus. The stimulus means external inputs to the network. (**a**) P = [−1 1 1 1 1]; T = [1], (**b**) P = [−1 −1 −1 −1 −1]; T = [1], (**c**) P = [1 −1 −1 −1 −1]; T = [−1], and (**d**) P = [1 −1 1 1 1]; T = [−1].

6. Comparison between Our Proposal and Related Works

Hardware design for neural architectures using base neurons built with CMOS transistors is a fundamental part of our design in comparison with the current cited works. Although we cannot compete in processing speed, we have the advantage of implementing ANN circuits with standard cells without modifying circuits for implementation. We chose several aspects to carry out a comparison among similar works [21–23] (see Table 8) considering: (1) the chosen material or device type, (2) the kind of device on which the synapsis is based, (3) the amount of power required, and (4) the learning algorithm used.

Table 8. Comparison between our proposal and related works.

Work	Device Quirks	Synapsis Based on	Power Consumption (mW)	Learning Algorithm
Our Proposal	CMOS	op-amps	46.08	BP
Zhang et al. [21]	Memristor	Memristor array + op-amps	-	BP
Krestinskaya et al. [22]	Memristive Crossbar	CMOS	115.65	BP
J. Han, et al. [23]	FPGA	LUT, FF, BRAM	477	SNNs

The synaptic weights of the proposed method, presented in Table 3, were obtained offline by the Backpropagation learning method for the analog implementations and by a Spiking Neural Network (SNN) for the field-programmable gate array (FPGA) implementation. In our proposal, the synaptic weights were implemented with resistors and op-amp circuits to form the base cell. In reference [21] the synapses were implemented with an array of resistances and op-amps. In contrast, in reference [22] the synapses were implemented by means of cells formed in the Memristive Crossbar. Finally, in the FPGA [23] they were implemented through modules that perform fixed arithmetic calculations using the hardware resources of the family.

A comparison between the proposed work and the implementation using FPGAs is somehow unfair because of the density of programmable logic of the latter, considering energy consumption, processing speed, and clock cycles, among other factors.

Power consumption is directly proportional to the density of the devices used to design the architectures at a hardware level. In the case of our proposal, power consumption is due just to op-amps and resistances. In the case of [22], power consumption is subject to the cells of the transistors, where the effect of signal propagation delay is not negligible. Finally, in the case of the FPGA implementation [23], power consumption was because of the density of programmable logic gates for the whole system; where we should consider static and dynamic powers. For instance, there is a device with unused hardware resources though consuming power since hardly an architecture will consume 100% of the FPGA resources, generating a non-optimal consumption of resources.

The architecture of [23] and our proposal are very different, in our case the architecture presents 9 neurons implemented in parallel and in the case of [23], the authors reported an architecture that calculated the 2842 neurons by using an iterative process (processor-type architecture), where a single neuron was implemented, and this implementation was able to compute the result of all neurons due to the design of the proposal. In this way, if the architecture of [23] implemented 9 neurons, as the same basic architecture requires, it would be consuming similar energy as they reported, since it was a processor architecture that was operating. Implementing 2842 neurons in the FPGA with floating numbers and their processing modules for addition, subtraction, etc., would require greater power consumption and hardware resources, which can be limited by the amount of FPGA resources.

We cannot compare the work in [22] and ours directly, even though both presented simulations carried out in Matlab and Pspice, respectively. For both cases, the technology and implementation were different. However, it is reasonable to think that the better the technology, such as memristors, the lower the energy consumption. In various works, such as in [24], it was said that the energy consumption of [22] was high, and they sought to lower it.

The work in [22] was developed in 180 nm TSMC's CMOS technology. The implementation and simulation of the circuit were carried out with memristors and in a crossbar configuration. The authors of [22] implemented a neural network with three neurons in the input layer, two neurons in the output layer, and five neurons in the hidden layer. In contrast, we proposed an ANN made with CMOS with 1.2 μm technology with five neurons in the input layer, one neuron in the output layer, and three neurons in the hidden layer.

Power dissipation was lower in our case because in the CMOS design, being this a mature technology, there are rules that must be satisfied [25–27]. On the other hand, in the case of memristors, their implementation in the commercial area is not available. Hence, the silicon organization of their components is still being built or defined. In the case of speed and performance, it is not competitive compared to DRAM, which is used by CMOS technology.

Other factors can affect the behavior of the circuits, but since they are different technologies, their usefulness in the comparison is null, so they are not part of the analysis of this work, such as frequency effects, transistor channel dimensions (W/L), channel resistance, channel conductivity, voltage threshold, response speed, scaling of CMOS devices, reduced geometry effects, channel length, and CMOS channel width.

7. Conclusions

In this work, an MLP architecture design at the hardware level was presented, using base neurons at the circuit level as its building blocks. Neuron weights and bias were optimized off-line using n-dimensional objective functions through the gradient descent method. The hardware architecture of the base neuron was designed on a CMOS operational amplifier in 1.2-micron technology. The total power consumption was 46 mW. The 5-3-1 ANN architecture was designed with nine neurons, using 36 op-amps, demonstrating its electronic performance in the practical case shown in Section 5.

The application developed with the implementation of the base neuron circuits showed optimal results in detecting the pattern. Starting with the supervised learning algorithm and based on the gradient descent method to obtain the synaptic weights and bias for the neural architecture, the responses shown in Figure 8 and Table 7 were appropriate to the electrical behavior. The MLP architecture presented an efficiency of 99 %. We might suggest other applications, but in the first instance, we determined that solving this problem was an adequate choice to show the efficiency of the MLP system. Other applications could be proposed, but that remains as future work; for example, we are currently working on image recognition to classify agricultural products and on data processing of hypertensive and diabetic patients. As part of the future work, we have the design and implementation of neural network circuits based on the optimization of objective functions. Besides, we can use other tools, such as Verilog-AMS or VHDL-AMS, as alternatives to simulate the results presented in this work. Another future work is trying other meta-heuristic optimization methods to obtain better ANN training results. We consider that the application of the proposed method was optimal but not without leaving a possibility of applying another meta-heuristic method in the short term.

The advantage of our development with respect to other works cited in Table 8, can be observed, mainly, in the following improvements: power dissipation, number of transistors, and probably, the dimension in area of the integrated circuit. It is likely that one of the disadvantages of our implementation is the speed of response, but that is compensated by the fact that this is a design containing just what is necessary for the application, without an excessive amount of resources.

Author Contributions: Conceptualization, A.M.-S., C.A.H.-G., L.A.M.-R. and I.A.-B.; methodology, A.M.-S., C.A.H.-G., L.A.M.-R. and I.A.-B.; software, A.M.-S., M.A.G. and J.A.O.T.; validation, A.M.-S., I.A.-B. and M.A.G.; formal analysis, A.M.-S. and L.A.M.-R.; investigation, A.M.-S., C.A.H.-G., L.A.M.-R. and I.A.-B.; resources, A.M.-S., C.A.H.G., L.A.M.-R., I.A.-B., M.A.G. and J.A.O.T.; writing—original draft preparation, A.M.-S., I.A.-B., L.A.M.-R. and C.A.H.-G.; writing—review and editing, A.M.-S., C.A.H.-G., L.A.M.-R. and I.A.-B.; visualization, C.A.H.-G. and J.A.O.T.; supervision, L.A.M.-R. and A.M.-S. All authors have read and agreed to the published version of the manuscript.

Funding: This research was funded by the Mexican National Council for Science and Technology (CONACYT) through the Research Projects 882, 278, and 613.

Institutional Review Board Statement: Not applicable.

Informed Consent Statement: Not applicable.

Data Availability Statement: Not applicable.

Conflicts of Interest: The authors declare no conflict of interest. The funders had no role in the study's design, in the collection, analyses, or interpretation of data, in the writing of the manuscript, or in the decision to publish the results.

References

1. Baker, R.J. *CMOS: Circuit Design, Layout, and Simulation*, 4th ed.; Wiley-IEEE Press: Hoboken, NJ, USA, 2019.
2. Razavi, B. *Design of Analog CMOS Integrated Circuits*, 2nd ed.; McGraw-Hill Education: New York, NY, USA, 2016; ISBN 0072524936.
3. Abbas, K. *Handbook of Digital CMOS Technology, Circuits, and Systems*; Springer International Publishing: Berlin/Heidelberg, Germany, 2020; ISBN 9783030371944.
4. Mandal, R.K. Design of a CMOS OR Gate using Artificial Neural Networks (ANNs). *AMSE J. Ser. Adv. D* **2016**, *21*, 66–77.
5. Miona, A.; Vančo, L. Application of Artificial Neural Networks in Electronics. *Electronics* **2017**, *21*, 87–94. [CrossRef]
6. Cichocki, A.; Unbehauen, R. *Neural Networks for Optimization and Signal Processing*; John Wiley & Sons, Inc.: Hoboken, NJ, USA, 1993.
7. Rizaner, F.B.; Rizaner, A. Approximate Solutions of Initial Value Problems for Ordinary Differential Equations Using Radial Basis Function Networks. *Neural Process. Lett.* **2018**, *48*, 1063–1071. [CrossRef]
8. Tan, L.S.; Zainuddin, Z.; Ong, P. Solving ordinary differential equations using neural networks. *AIP Conf. Proc.* **2018**, *1974*, 020070.
9. Pates, R.; Vinnicombe, G. Scalable Design of Heterogeneous Networks. *IEEE Trans. Autom. Control* **2017**, *62*, 2318–2333. [CrossRef]
10. Haug, Y.; Guo, N.; Seok, M.; Tsividis, Y.; Mandli, K.; Sethumadhavan, S. Hybrid Analog-Digital Solution of Nonlinear Partial Differential Equations. In Proceedings of the 2017 50th Annual IEEE/ACM International Symposium on Microarchitecture (MICRO), Cambridge, MA, USA, 14–18 October 2017. [CrossRef]
11. Hagan, M.T.; Demuth, H.B.; Beale, M.H. *Neural Network Design*, 2nd ed.; PWS Publishing Company: Boston, MA, USA, 2014.
12. Adolphs, L.; Daneshmand, H.; Lucchi, A.; Hofmann, T. Local Saddle Point Optimization: A Curvature Exploitation Approach. Proceedings of Machine Learning Research. In Proceedings of the 22nd International Conference on Artificial Intelligence and Statistics, Naha, Japan, 16–18 April 2019.
13. Rhinehart, R.R. *Engineering Optimization: Applications, Methods and Analysis*; John Wiley & Sons: Hoboken, NJ, USA, 2018.
14. French, M. *Fundamentals of Optimization Methods, Minimum Principles and Applications for Making Things Better*; Springer: Berlin/Heidelberg, Germany, 2018.
15. Bogdan, M.; Wilamowski, J.; David, I. *Intelligent Systems*; CRC Press: Boca Raton, FL, USA, 2018. [CrossRef]
16. Li, Y. *Deep Reinforcement Learning: An Overview, Neural Networks*; Elsevier: Amsterdam, The Netherlands, 2018.
17. Markopoulos, A.P.; Georgiopoulos, S.; Manolakos, D.E. On the use of back propagation and radial basis function neural networks in surface roughness prediction. *J. Ind. Eng. Int.* **2016**, *12*, 389–400. [CrossRef]
18. Lai, J.L.; Wu, C.Y. Design Ratio-Memory Cellular Neural Network (RMCNN) in CMOS Circuit Used in Association-Memory Applications for 0.25 mm Silicon Technology. *Open Mater. Sci. J.* **2016**, *10*, 54–69. [CrossRef]
19. Su, Z.; Kolbusz, J.; Wilamowski, B.M. Linearization of Bipolar Amplifier Based on Neural-Network Training Algorithm. *IEEE Trans. Ind. Electron.* **2016**, *63*, 3737–3744. [CrossRef]
20. Burr, G.W.; Shelby, R.M.; Sebastian, A.; Kim, S.; Kim, S.; Sidler, S.; Virwani, K.; Ishii, M.; Narayanan, P.; Fumarola, A.; et al. Neuromorphic computing using non-volatile memory. *Adv. Phys. X* **2017**, *2*, 89–124. [CrossRef]
21. Zhang, Y.; Wang, X.; Friedman, E.G. Memristor-Based Circuit Design for Multilayer Neural Networks. *IEEE Trans. Circuits Syst. Regul. Pap.* **2018**, *65*, 677–686. [CrossRef]
22. Krestinskaya, O.; Salama, K.N.; James, A.P. Analog Backpropagation Learning Circuits for Memristive Crossbar Neural Networks. In Proceedings of the 2018 IEEE International Symposium on Circuits and Systems (ISCAS), Florence, Italy, 27–30 May 2018; pp. 1–5.
23. Han, K.J.; Li, Z.; Zheng, W.; Zhang, Y. Hardware Implementation of Spiking Neural Networks on FPGA. *Tsinghua Sci. Technol.* **2020**, *25*, 479–486. [CrossRef]
24. Nafea, S.F.; Dessouki, A.A.S.; El-Rabaie, S. Summary of memristor up to 2015. *Menoufia J. Electron. Eng. Res.* **2015**, *24*, 79–106. [CrossRef]
25. Wang, Y.; Cao, S.; Jin, X.; Peng, Y.; Luo, J. *Semiconductor Science and Technology*; IOP Publishing Ltd.: Bristol, UK, 2020; Volume 36.
26. Yener, Ş.; Kuntman, H. New CMOS based memristor implementation. In Proceedings of the 2012 International Conference on Applied Electronics, Pilsen, Czech Republic, 5–7 September 2012.
27. Abunahla, H.; Mohammad, B. *Memristor Technology: Synthesis and Modeling for Sensing and Security Applications, Chapter 1 Memristor Device Overview and Chapter 6 Memristor Device Modeling*; Springer: Berlin/Heidelberg, Germany, 2018.

 *sensors*

Article

Excel Methods to Design and Validate in Microelectronics (Complementary Metal–Oxide–Semiconductor, CMOS) for Biomedical Instrumentation Application

Graciano Dieck-Assad [1], José Manuel Rodríguez-Delgado [1] and Omar Israel González Peña [1,2,*]

[1] Tecnológico de Monterrey, School of Engineering and Science, Av. Eugenio Garza Sada Sur No. 2501, col. Tecnológico, Monterrey 64849, Mexico; graciano.dieck.assad@tec.mx (G.D.-A.); jmrd@tec.mx (J.M.R.-D.)
[2] Tecnológico de Monterrey, Institute for the Future of Education, Av. Eugenio Garza Sada Sur No. 2501, col. Tecnológico, Monterrey 64849, Mexico
* Correspondence: ogonzalez.pena@gmail.mx

Citation: Dieck-Assad, G.; Rodríguez-Delgado, J.M.; González Peña, O.I. Excel Methods to Design and Validate in Microelectronics (Complementary Metal–Oxide–Semiconductor, CMOS) for Biomedical Instrumentation Application. *Sensors* **2021**, *21*, 7486. https://doi.org/10.3390/s21227486

Academic Editors: Jong-Ryul Yang and Seong-Tae Han

Received: 25 September 2021
Accepted: 30 October 2021
Published: 11 November 2021

Publisher's Note: MDPI stays neutral with regard to jurisdictional claims in published maps and institutional affiliations.

Abstract: CMOS microelectronics design has evolved tremendously during the last two decades. The evolution of CMOS devices to short channel designs where the feature size is below 1000 nm brings a great deal of uncertainty in the way the microelectronics design cycle is completed. After the conceptual idea, developing a thinking model to understand the operation of the device requires a good "ballpark" evaluation of transistor sizes, decision making, and assumptions to fulfill the specifications. This design process has iterations to meet specifications that exceed in number of the available degrees of freedom to maneuver the design. Once the thinking model is developed, the simulation validation follows to test if the design has a good possibility of delivering a successful prototype. If the simulation provides a good match between specifications and results, then the layout is developed. This paper shows a useful open science strategy, using the Excel software, to develop CMOS microelectronics hand calculations to verify a design, before performing the computer simulation and layout of CMOS analog integrated circuits. The full methodology is described to develop designs of passive components, as well as CMOS amplifiers. The methods are used in teaching CMOS microelectronics to students of electronic engineering with industrial partner participation. This paper describes an exhaustive example of a low-voltage operational transconductance amplifier (OTA) design which is used to design an instrumentation amplifier. Finally, a test is performed using this instrumentation amplifier to implement a front-end signal conditioning device for CMOS-MEMS biomedical applications.

Keywords: freeware; open science; analog microelectronics design; long channel transistors; short channel transistors; integrated circuit design; CMOS design; VLSI; higher education; educational innovation; integrated circuit layout; complex thinking

1. Introduction

Currently, we have high-capacity technology of analog and digital electronic devices due to the micro-components that are increasingly becoming smaller in scale. In fact, in 1965, Gordon E. Moore predicted (Moore's law) that every 2 years the number of transistors in a microprocessor would double. This law worked for the first 10 years [1]; then it became a joke between engineers who said that "now people predict that the end of Moore's law doubles every 2 years." Since then, transistor integration has been observed as illustrated in Figure 1, where Moore's law was in force for a long time. However, the limits towards a scale in nanotechnology have shown that although Moore's law no longer applies, significant efforts are still being made to continue reducing the size of the micro and nano components. For example, recently design and process in semiconductors have been done with the development of the world's first chip with 2 nanometers (nm) nanosheet technology [2].

131

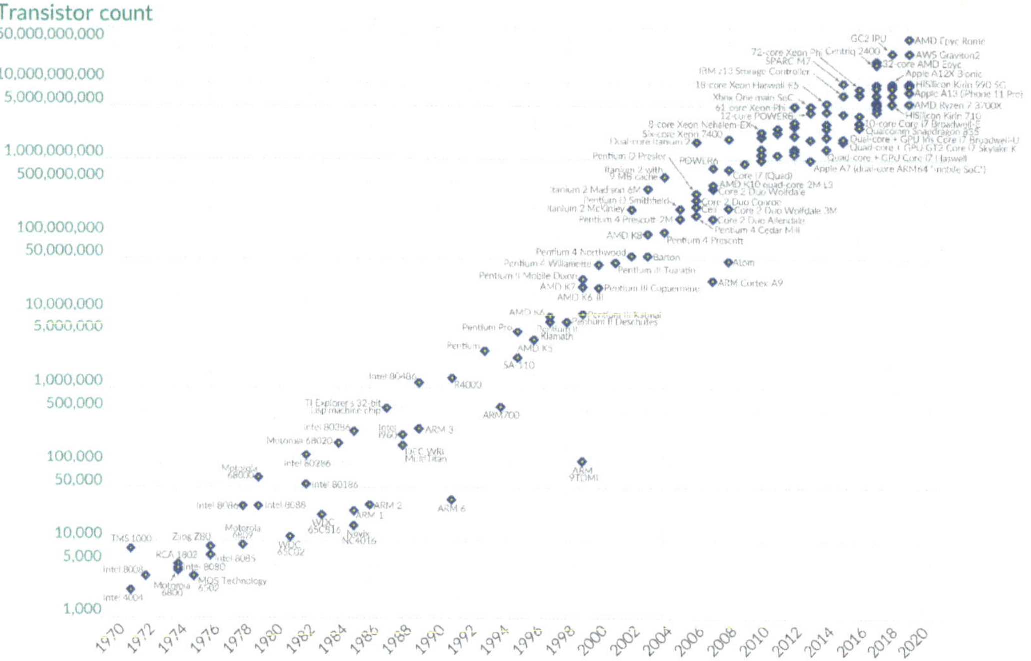

Figure 1. Behavior of the technological development for the small-scale design of transistors on microchips doubling every two years (Moore's law) [1].

Therefore, with this context in which the microchip technology is reducing in size over time, college students with different science and engineering majors need to understand the "know-how" of designing circuits that integrate microchips. Unfortunately, often this knowledge is taught with specialized complex (and expensive) software, which is sometimes not readily available in many universities around the world.

Analog and digital integrated circuit design has emphasized a good understanding of analog and digital electronic circuits to model, simplify, analyze, and simulate microelectronic devices before developing layouts and sending devices to IC fabrication foundry consortiums. Design engineers find SPICE simulations useful to validate "thinking models" [3,4] to verify cause–effect relationships and to ensure that assumptions made are appropriate in the design process. Microelectronics courses and specialized workshops (and webinars) teach conceptual design where the following concepts are emphasized.

- The importance of the integrated circuit (IC) design in microelectronics engineering is embedded into electronic engineering innovation and design flow.
- Analog CMOS microelectronics focus on the electrical and physical design processes.

IC design is important in device development for microelectronics engineering innovation and is embedded into the design flow which may have continuous iterations to optimize designs by decision refinements. In the process of CMOS microelectronics conceptual design [5–7], the electrical device specification requires active and passive models for creating, verifying, and determining the robustness of the design. This process involves the selection of a conceptual circuit, the analysis of the selected circuit, the possibility of a modification to the circuit, and the verification of the circuit solution. The physical electronics design process [8–10] consists of representing the electrical device in a 2D layout consisting of many different geometrical rectangles at various levels (layers). This layout is then used to implement a 3D integrated circuit during the fabrication process. The device conceptual model follows a process to obtain a layout. This process includes [3,4]:

1. The W/L (width/length of transistors) values and schematic (usually from SPICE simulations).
2. A computer-aided tool (CAD) system is used to enter geometries.
3. The engineer must obey a set of rules called design rules. These rules establish the fabrication limitations and ensure that the device is robust and reliable.
4. Once the layout is complete, a process called layout versus schematic (LVS) is applied to determine if the physical layout represents the electrical schematic.
5. Parasitic elements are extracted, once the physical dimensions of the designs are known. They include capacitances from conductor to ground or between conductors and bulk resistances.
6. The parasitic elements are entered into the simulation database and the design is re-stimulated to ensure they will not cause a device failure.

This process is depicted in the flow diagram shown below in Figure 2.

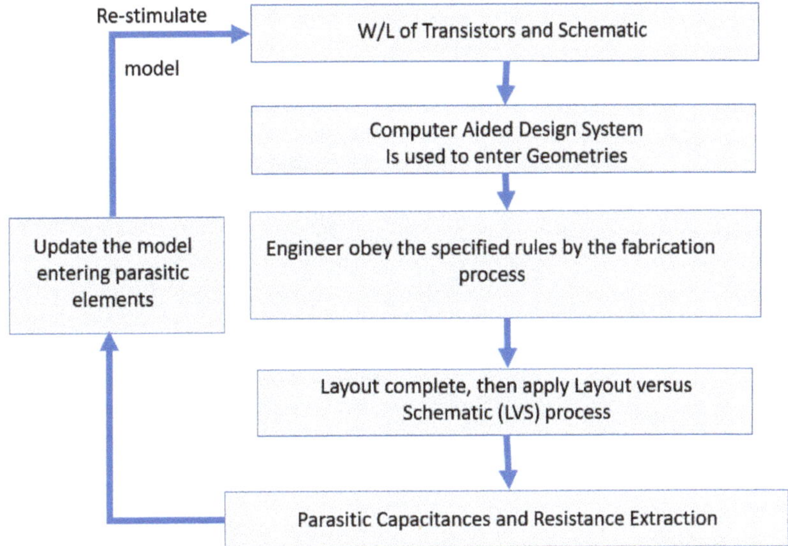

Figure 2. Block diagram showing the design procedure to reach the Layout conceptual model.

Once the first approach to design is terminated, the process continues with design testing which consists of coordinating, planning, and implementing the measurement of the integrated circuit performance. The objective is to compare the experimental performance with the specifications and/or simulation results. Several tests are available, e.g., functional: verification of the nominal specifications; parametric: verification of the characteristics to within a specified tolerance, verification of the static (AC and DC) characteristics of the circuit or system, and verification of the dynamic (transient) characteristics of a circuit or system. Additional testing could include device testing performed at the wafer level or package level and detailed testing that removes the influence of measurement system in the device performance.

The conceptual design of analog integrated circuits in new devices is now shaped by rules such as: consumers generate a need for new integrated circuits, design engineers have an open possibility of participating in designs, time to develop a product is reduced, profit in products and prototypes are not readily necessary, and the new concept of "crowd designing" [3] plays an important role in device development. This article discusses Excel methods to perform "paper and pencil" calculations ("thinking model") in the conceptual design of CMOS analog integrated circuits which are validated using ELECTRIC-LTSpice to provide the initial characteristics of the device. Several methodologies have been developed for CMOS device design and testing but most of them are very specific to the final prototype or experimental application [11–17]. The Excel methods analyzed in this study focus on providing didactic instruction in microelectronics for undergraduate and graduate students. Therefore, it seeks to focus learning by referring to world trends in conducting open science, providing the social appropriation of knowledge. The forefront is didactic management whose central axis is the catalyst of open innovation processes that have proven to be very successful disruptive models in open laboratories, universities, research centers, industry, and government for the development of emerging economies and public policies [18–26].

As a result, this study describes a method that uses an Excel spreadsheet to start the conceptual design from the point of view of the "thinking model" or the first cut evaluation of the design with "paper and pencil". In addition, several contributions stand out:

The Excel methods discussed here focus on microelectronics education for undergraduate and graduate students. The article describes a simple method using an Excel spreadsheet to initiate conceptual design from the standpoint of "thinking model" or "paper and pencil" first cut evaluation of the design. Furthermore, several contributions are emphasized:

(a) This work describes both the conceptual design evaluations using the traditional equations and prepares the way for the layout implementation by setting up the schematic of the design.

(b) In addition, this study describes that once the schematic simulation agrees with the scheduled specifications and the "thinking model" provides a possible design solution, the layout is developed and a comparison between the output results with the thinking model specs validates the third phase of the process.

(c) Moreover, the methodology applied in this study can be developed for more complicated CMOS analog integrated circuits (IC) conceptual designs, specifically for teaching purposes.

(d) Complex conceptual designs can also be addressed using simple and readily available software (freeware) to teach with an open science view.

(e) Furthermore, this article illustrates an exhaustive example of a low voltage operational transconductance amplifier (OTA) design for portable biomedical applications. The test is performed using this instrumentation amplifier to implement a front-end signal conditioning device for CMOS-MEMS biomedical applications.

This manuscript is organized as follows. Section 2 briefly describes device modeling to represent transistors using first principle equations that predict their behavior in different regimes and operating conditions. Section 3 provides the basic MOSFET modeling equations, starting with the threshold voltage calculations, the transconductance equations for the ohmic (sub-saturation), active (saturation), and subthreshold conditions, both for long channel and short channel devices. Section 3 also gives the typical MOSFET parameters for 0.5 mm CMOS technology, the small-signal model parameter equations, and useful resistance and capacitance calculations using the foundry process parameters for the technology. Section 4 describes the Excel methods that students use to develop their first cut approximations in conceptual designs of CMOS devices and before testing schematic simulations and layouts. This section explains three Excel methodologies: single straight, tabular straight, and two-dimensional processing methods to perform the evaluation of the device's conceptual design. This section also discusses resistance, capacitance, and differential amplifier conceptual designs as specific examples to apply the Excel methods. Section 5 discusses the Excel methods applied for complete amplifier design. Here the cascode amplifier and the OTA (Operational Transconductance Amplifier) are used as examples of how students use the basic two-dimensional Excel methodology to solve CMOS microelectronic conceptual design devices. Section 6 illustrates a complete case study, developed by students and wrapped up by their instructor, of a low-power high-gain operational amplifier for biomedical applications. In this section, comparison of the design is performed with respect to a particular device appearing in the technical literature and

an application of a CMOS-MEMS signal conditioning is developed considering the requirement of conditioning a differential mode temperature sensor´s output to obtain a readily
available low voltage representation of the temperature of a micro hotplate multisensor
platform. Finally, Section 7 wraps up the paper with illustrative conclusions about how
these Excel methods have provided extraordinary insights to undergraduate students in
their effort to consolidate a good understanding of microelectronics conceptual integrated
circuit (IC) design.

2. Materials and Methods—Device Modeling

The process of device modeling consists of representing the electrical properties of devices using mathematical equations, circuits, graphs, correlations, and energy conservation
laws. Models allow predictions and validation of circuit performance with uncertainties
coming from no-idealities and non-linear behaviors in electronic components. Typical
equations and conservation laws are Ohm´s law, large- and small-signal models of MOS-
FET transistors, VTC curves, and I-V curves of diodes. The final goals are to simplify the
cause–effect relationships allowing the engineer to understand and consider decisions that
increase performance in the circuit.

Analog integrated circuits in microelectronic conceptual design are developed using a
non-hierarchical structure where the use of repeated blocks is only possible in few devices,
and therefore the design process is complex and challenging. To handle this, design
engineers use hierarchy whenever possible, use good organization techniques, efficiently
document the design, provide reasonable and reliable assumptions and simplifications, and
eventually validate the conceptual designs using simulation experiments. Assumptions
and simplifications are used to emphasize the essential characteristics by neglecting the non-
dominant effects in the design. The challenge of teaching microelectronics is to develop
an insight into the design process without requiring specialized professional software
which is not readily available in many universities around the world. Figure 3 shows
the microelectronics design process, on the left side, with the insight given on the right
side, using Excel methods and other simulation techniques which are readily available
to universities.

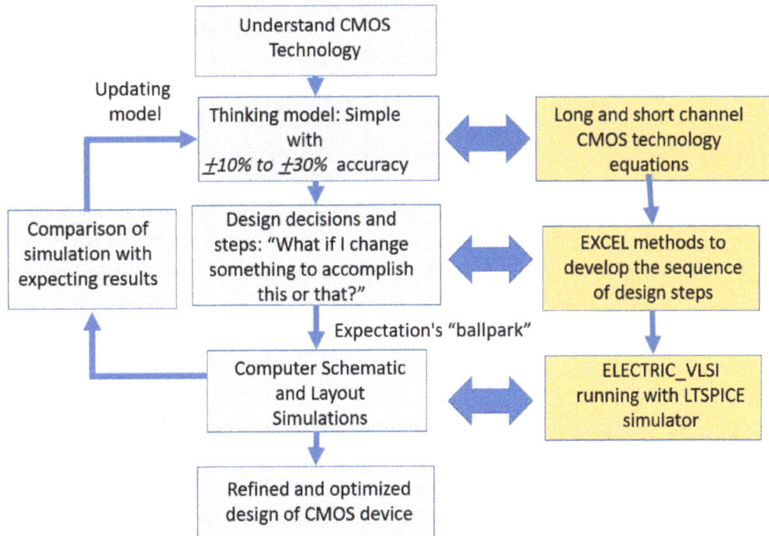

Figure 3. The CMOS microelectronics design process and the insight provided by the Excel methods
in the quest to teach microelectronics design with simple tools.

Analog integrated circuit design and device evaluation have reached a level of maturity in established applications such as digital to analog and analog to digital conversion systems, front end signal conditioning devices, instrumentation channel devices, bandgap reference sources, DC to DC power conversion drives, and other important microelectronic circuits [3]. Finally, analog circuit conceptual designs have significant applications in devices where speed and power have an overwhelming advantage over digital devices. This paper reviews the long and short channel models for CMOS devices and further application of long channel equations using Excel methods for first cut approximations before computer simulations and layout development. Those approximation models can be found in many CMOS microelectronics specialized textbooks [3,4,7,27–29] and they are summarized here for completeness in this discussion.

3. Results—Transistor Models for Analog Conceptual Design

3.1. Threshold Voltage Calculation

The evaluation of threshold voltage, V_T, is fundamental in the development of CMOS microtechnology because gives the necessary condition for allowing operation of the transistors in the right zone.

$$V_T = [V_{T0}] + \gamma \left(\sqrt{|2\varphi_F| + V_{SB}} - \sqrt{|2\varphi_F|} \right) \tag{1}$$

where γ is the body factor, $2\varphi_F$ is the Fermi potential, V_{T0} is the zero bias threshold voltage, and V_{SB} is the potential difference between source and bulk of the device.

3.2. Current Equations for Long Channel Devices

The drain-source current equations for the MOSFET long channel device are given below for $V_{GS} \geq V_T$ and $k = \mu C_{ox}(W/L)$:

Saturation zone: $V_{DS} \geq V_{GS} - V_T$

$$i_D = \frac{k}{2}[V_{GS} - V_T]^2(1 + \lambda V_{DS}) \tag{2}$$

Ohmic zone: $V_{DS} \leq V_{GS} - V_T$

$$i_D = \frac{k}{2}[2(V_{GS} - V_T) - V_{DS}]V_{DS} \tag{3}$$

For $V_{GS} \leq V_T$ the transistor plays in the subthreshold zone with the following exponential behavior, like the bipolar junction transistor:

$$i_D = I_t \frac{W}{L} e^{\frac{(V_{GS} - V_T)}{nV_t}} \tag{4}$$

where the nomenclature and parameters of the model are as follows: i_D is the drain to source current flowing in the MOSFET, μ is the charge carrier mobility, C_{ox} is the SiO_2 oxide capacitance, (W/L) is the width/length transistor ratio, V_{GS} is the gate to source potential differences, V_{DS} is the drain to source potential differences, λ is the transistor´s channel modulation parameter, I_t is the subthreshold saturation current (<1 µA), V_t is the thermal voltage (~26 mV at 25 °C), and n is a subthreshold constant (~1 to 2).

3.3. Velocity Saturation and Effective Mobility in Current Equations of Short Channel Devices

To consider velocity saturation and effective mobility in short channel MOSFETs adjustments are made in the model equations presented above. This is particularly important in the saturated and ohmic regions of operation. The drain-source current equations for the MOSFET short channel device are given below for $V_{GS} \geq V_T$, having saturation velocity $v_{sat} = \mu_e E_c/2$, where μ_e is now the effective mobility of charge carriers and E_c is the critical field for which the carrier saturation occurs.

Saturation zone:

$$V_{DS}(sat) = \frac{(V_{GS} - V_T)E_c L}{(V_{GS} - V_T) + E_c L} \tag{5}$$

$$i_D = \frac{W\mu_e C_{ox}E_c}{2}\left[\frac{(V_{GS} - V_T)^2}{(V_{GS} - V_T) + E_c L}\right] \tag{6}$$

Ohmic zone:

$$V_{DS}(ohmic - limit) = V_{DS}(sat) = \frac{(V_{GS} - V_T)E_c L}{(V_{GS} - V_T) + E_c L} \tag{7}$$

$$i_D = \frac{W\mu_e C_{ox}E_c}{(E_c L + V_{DS})}\left[(V_{GS} - V_T)V_{DS} - \frac{V_{DS}}{2}\right]V_{DS} \tag{8}$$

If we summarize the SPICE-level I models of the MOSFET transistors to be used in conceptual CMOS designs, Table 1 describes the equations considering the ohmic and saturation regions in the so-called strong inversion regime.

Table 1. Summary of CMOS N-Channel transistor large signal model.

	Long Channel L > 1 μm	**Short Channel L < 1 μm**
Ohmic	$i_{DS} = \frac{k'W}{L}\left[(V_{GS} - V_T) - \frac{V_{DS}}{2}\right]V_{DS}$	$i_{DS} = \frac{\mu_e C_{ox}E_c W}{(E_c L + V_{DS})}\left[(V_{GS} - V_T)V_{DS} - \frac{V_{DS}}{2}\right]V_{DS}$
$V_{DS}(sat)$	$V_{DS}(sat) = V_{GS} - V_T$	$V_{DS}(sat) = \frac{(V_{GS} - V_T)E_c L}{(V_{GS} - V_T) + E_c L}$
Saturation	$i_{DS} = \frac{kW}{2L}[V_{GS} - V_T]^2$	$i_D = \frac{W\mu_e C_{ox}E_c}{2}\left[\frac{(V_{GS} - V_T)^2}{(V_{GS} - V_T) + E_c L}\right]$
Notes:		$k' = \mu_0 C_{ox}$ E_c = critical horizontal field E_y μ_e = carrier mobility considering horizontal field E_y

Moreover, for "paper and pencil" calculations ("Thinking Modeling") the parameters used for 0.5 μm CMOS technology are described in Table 2.

Table 2. MOSFET parameters for CMOS 0.5 μm process (C5_models) obtained in ON-SEMI Wafer Runs [6].

Parameter Symbol	Description	Parameter Value		
		N-Channel	P-Channel	Units
V_{T0}	Threshold voltage ($V_{bs} = 0$)	0.76	−0.96	V
K'	Transconductance Parameter in saturation	115.6	37.8	μA/V^2
γ	Bulk threshold parameter	0.49	0.56	V$^{1/2}$
λ	Channel Modulation parameter	0.04	0.05	V^{-1}

3.4. Small-Signal Model of Devices

The useful small-signal equations for MOSFET transistors correspond to the saturation zone of the device. The two most important parameters are g_m, transconductance gain, and g_{ds}, output conductance of the device which provides its output resistance [3,4]. The equations for those parameters are described as follows.

$$g_m = \left.\frac{di_{DS}}{dV_{GS}}\right|_Q = \sqrt{2\beta I_D(1 + \lambda V_{DS})} \cong \sqrt{2\beta I_{DS}} \tag{9}$$

$$g_{ds} = \left.\frac{di_{DS}}{dV_{DS}}\right|_Q = \frac{\lambda I_{DS}}{1 + \lambda V_{DS}} \cong \lambda I_{DS} \tag{10}$$

From those expressions, λ is the channel modulation parameter shown in Table 2 and $\beta = k = \mu Cox(W/L)$ which was defined earlier for the large-signal model. These parameters are evaluated at the quiescent Q point where the device operates and the small-signal ignites.

3.5. Parasitic Capacitances

The most important parasitic elements of the MOSFET are the capacitances due to the inherent field-effect operation of this transistor. The oxide and p-n junction capacitances are responsible for limitations of the frequency response which generates poles and zeros which are frequently analyzed to determine the stability of the device and the dominant bandwidth of the system. Three capacitances are considered in the conceptual design modeling:

i. "Oxide capacitance" is formed by the SiO_2 between the gate terminal and the channel formed from drain to source. This capacitance is given by the gate oxide capacitance as follows.

$$C_g = WLC_{ox} = WL\epsilon_{ox}/t_{ox} \tag{11}$$

where ε_{ox} is the oxide dielectric constant and t_{ox} is the oxide thickness.

ii. "P-N junction capacitances" due to the reverse bias of the p-n junctions formed between drain/source to the substrate/bulk terminals of the device. This capacitance is evaluated using the built-in potential φ_B, built-in zero-bias junction capacitance C_{j0}, and the built-in junction capacitance C_j as follows.

$$\varphi_B = V_t ln\left(\frac{N_A N_D}{n_i^2}\right) \tag{12}$$

$$C_{j0} = \sqrt{\frac{\epsilon_{Si}qN_A}{2\varphi_B}} \tag{13}$$

$$C_j = \frac{C_{j0}}{\left(1 - \frac{V_j}{\varphi_B}\right)^m} \tag{14}$$

where n_i is the intrinsic carrier concentration in silicon, N_A/N_D are numbers of acceptor/donor atom concentrations per volume in material, ε_{Si} is the silicon dielectric constant, and V_j is the reverse bias voltage applied.

3.6. Passive Components

In microelectronic conceptual design, the passive components provide additional versatility in the consolidation and refinement of many integrated circuits that require compensation, parasitic element adjustment, antenna integration, and other possible maneuvers. For the conceptual modeling using the Excel methodologies, this paper considers only resistor and capacitor implementations. To implement resistances, Table 3 shows the values for sheet and contact resistances in the C5 ON-SEMI process [15] described by wafer runs from the corresponding foundry.

Table 3. Resistance process parameters for CMOS 0.5 μm process used in ON-SEMI wafer runs [13].

Process Parameters	N+	P+	POLY	POLY2	POLY2 HR	N_{well}	M1	M2	M3	UNITS
Sheet resistance	82.4	106.7	23.2	40.8	1076	808	0.09	0.09	0.05	Ω/sq
Contact resistance	59.6	152.5	16	26	-	-	0.84	0.84	0.82	Ω

From Table 3, N+ and P+ are the n-type and p-type doped active silicon materials, respectively; POLY, POLY2, and POLY2_HR are polysilicon, polysilicon-2, and polysilicon-

2-high-resistivity materials, respectively; N_{WELL} corresponds to the n-type well (where P-Type transistors are located); M1, M2, and M3 are metal-1, metal-2, and metal-3 layers, respectively, where the component connections are implemented inside the chip.

To implement capacitances, Table 4 shows the values for area, fringe and overlap capacitances in C5 ON-SEMI process [15]. The area capacitances for substrate, N+ active, P+ active, POLY, POLY2, M1 and M2 have units of aF/μ^2, while the fringe and overlap capacitances are given in units of aF/μ. Parameters from table IV are very important in the compensation of analog integrated circuits that are used in instrumentation channels for processing signals coming from sensors and transducers.

Table 4. Capacitance process parameters for CMOS 0.5 μm process used in ON-SEMI wafer runs [15].

Capacitance Parameters	N+	P+	POLY	POLY2	M1	M2	M3	N_{well}	UNITS
Area(substrate)	416	710	86	-	29	12	8	91	aF/μm^2
Area(N+ active)	-	-	2456	-	-	-	-	-	aF/μm^2
Area(P+ active)	-	-	2456	-	-	-	-	-	aF/μm^2
Area(POLY)	-	-	-	922	64	16	9	-	aF/μm^2
Area(POLY2)	-	-	-	-	58	-	-	-	aF/μm^2
Area(M1)	-	-	-	-	-	32	12	-	aF/μm^2
Area(M2)	-	-	-	-	-	-	32	-	aF/μm^2
Fringe(substrate)	345	236	-	-	51	34	26	-	aF/μm^2
Fringe(POLY)	-	-	-	-	70	39	28	-	aF/μm^2
Fringe(M1)	-	-	-	-	-	49	33	-	aF/μm^2
Fringe(M2)	-	-	-	-	-	-	55	-	aF/μm^2
Overlap(N+active)	-	-	191	-	-	-	-	-	aF/μm^2
Overlap(P+active)	-	-	234	-	-	-	-	-	aF/μm^2

The equations to implement passive components are given as follows. Resistances made in N_{well}, N+, P+, POLY, POLY2 and POLY2_HR:

$$R_T = [Sheet\ Resistance\ Parameter] \times L/W \tag{15}$$

Capacitances made by POLY-POLY or POLY-Metal:

$$C_T = [Capacitance\ Parameter] \times WL \tag{16}$$

Now that the model equations have been presented, the following sections will discuss the methodology used to describe the conceptual model for the CMOS microelectronic circuit.

4. Excel Methods for CMOS Design

Conceptual design involves the use of parameters and constants to compute and size values of components and electrical variables [16,17]. However, sometimes the number of specifications given is larger than the number of degrees of freedom available to the designer trying to comply and fulfill them [3,4,27–29]. Therefore, in the design process, a series of decision points is necessary to iterate the processing flow before obtaining the final solution. Several trials are necessary and sometimes, intermediate simulations are required to analyze different alternative solutions. In this case, the use of Excel methods is convenient while advancing in the design process. There are three sorts of recommended methods:

a. Single straight processing.
b. Tabular straight processing.
c. Two-dimensional processing.

4.1. Single Straight Processing Excel Method

In the single straight processing, the method proceeds in a horizontal fashion with the CMOS technology parameters in the first block of columns. Then, the processing flow continues sequentially, column by column, until the final desired calculation values are obtained. If several conditions are considered, then a row repetition is developed accordingly. Figure 4 shows single straight processing to design resistances using active p+, active n+, polysilicon and polysilicon-2 CMOS elements with their respective contact resistances and varying W and L sizes. Figures 5 and 6 show the CMOS 500 nm IC layouts of 3.5 K and 1.5 K resistors, obtained from Electric-VLSI. The artwork was developed once the Excel method was used to calculate the number of sheets or squares required by the conceptual design.

Processing Flow

Resistance parameters			Contact Resistance					Equivalent Resistance		
Ohms per square			Ohms					p+	n+	
p+	n+		p+	n+		W	L	Req p+	Req n+	
106.7	82.4		152.5	59.6		3	95	3531.3	2668.9	
Ohms per square			Ohms					Equivalent Resistance		
Poly	Poly-2		Poly	Poly-2		W	L	Poly	Poly-2	
23.2	40.8		16	26		5	334	1565.8	2751.4	

Figure 4. Single straight method to design CMOS resistances.

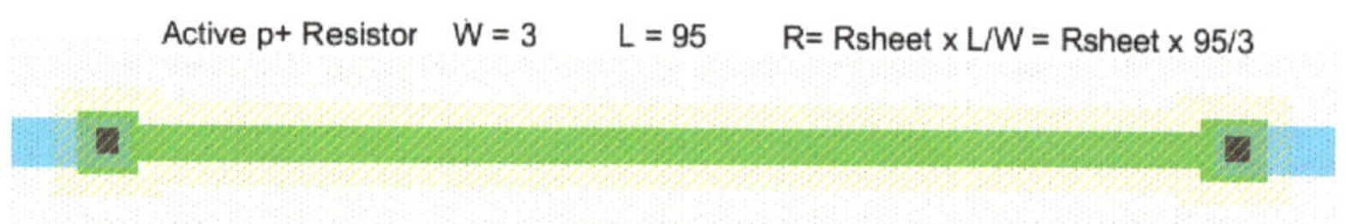

Figure 5. CMOS IC layout active p+ resistor design, 3.5 K, from Electric-VLSI.

Figure 6. CMOS IC layout polysilicon resistor, 1.5 K, from Electric-VLSI.

4.2. Tabular Straight Processing Excel Method

In the tabular processing, the method proceeds as usual, in a horizontal fashion to evaluate the design instance with the CMOS technology parameters in the first block of columns. However, this evaluation is repeated for a multiple number of instances while varying one or two specifications or parameters. Then, the processing flow continues sequentially, row by row, until the final desired instance is terminated. Each instance considered will be evaluated over a single row. Figure 7 and Figure 8 show tabular processing to design capacitances using poly-poly-2 CMOS 500 nm technology. Figure 5 starts with the value of the capacitance and ends up with squared plates in Poly-Poly to design the capacitance. Figure 7 starts with L/2 feature size (300 nm in this case) X-Y rectangular dimensions to obtain the effective capacitance value. Figures 9 and 10 show the CMOS 500 nm IC layouts of 2 pF and 10 pF capacitances, obtained from Electric-VLSI. The artwork was also developed once the Excel methodology was used to evaluate the conceptual design. Capacitors from Figures 8 and 9 are used for operational transconductance amplifier (OTA) compensation later in this paper.

Processing Flow

From Cp-p to X-Y dimensions					Dimensions 1x1		Dimensions 2x1		Parasitic
Req	$C_{poly2\text{-}poly}$	C5-Mod	Cpoly-sub	Estimated					
Cp-p	Poly2-Poly	λ	Poly-Sub	Area sq($\mu\mu$)	X	Y	X	Y	Cp-s
pF	aF/μm^2	μm	aF/μm^2	μm^2	λs	λs	λs	λs	fF
0.4	922	0.3	86	4820.44	69.4	69.4	138.9	34.7	37.31
0.5	922	0.3	86	6025.55	77.6	77.6	155.2	38.8	46.64
1	922	0.3	86	12051.10	109.8	109.8	219.6	54.9	93.28
1.5	922	0.3	86	18076.64	134.4	134.4	268.9	67.2	139.91
2	922	0.3	86	24102.19	155.2	155.2	310.5	77.6	186.55
3	922	0.3	86	36153.29	190.1	190.1	380.3	95.1	279.83
5	922	0.3	86	60255.48	245.5	245.5	490.9	122.7	466.38
7.5	922	0.3	86	90383.22	300.6	300.6	601.3	150.3	699.57
10	922	0.3	86	120510.97	347.1	347.1	694.3	173.6	932.75
	.λ= SCALE SIZE OF THE ELECTRIC_VLSI LAYOUT PROGRAM								

Figure 7. Tabular method to design CMOS capacitances. This Excel goes from capacitance value to X-Y dimensions.

4.3. Two-Dimensional Processing Excel Method

In two-dimensional processing, the method proceeds with the horizontal first row having the specifications and CMOS technology parameters of the conceptual design. Each row will define a step in the processing design flow such that the Excel will progress down and away from the first cell of the spreadsheet. The evaluations are arranged such that intermediate calculations follow a slope down from the early decisions all the way to the last decision. Usually, an iterative process is necessary to comply with two or three specifications with a single degree of freedom. For instance, with the bias current, I_{SS}, of a differential amplifier, can fulfill expectations for Slew Rate (SR), Output Resistance (R_{out}), and Power dissipation (P_{diss}) in the conceptual design of a CMOS differential amplifier. Figure 11 illustrates the Excel method to develop the conceptual design for a CMOS differential amplifier. The method illustrates the step-by-step evaluation and decision-making process downward, and the CMOS technology specifications are shown rightwards as shown in Figure 11. The CMOS Technology characteristics flow horizontally to the right and the CMOS design equations, from (5) to (10), flow downwards illustrating the step-by-by step procedure. The evaluation of CMOS transistor size, W/L, goes along to fulfill the required specification. However, if a particular spec does not convince the design engineer, the processing flow can be stopped, and a recalculation with a different spec or different decision making is readily possible at every row. The processing flow continues row/column by row/column until the final step and size selection is terminated.

Processing Flow

From X-Y dimensions to Cp-p							
Dimensions		Cpoly2-poly	C5-mod	C from C5	Estimated	Equivalent	Parasitic
X	Y	Poly2-Poly	λ	Poly-Sub	Area sq($\mu\mu$)	Cp-p	Cp-s
λs	λs	aF/μm^2	μm	aF/μm^2	μm^2	pF	fF
68	68	922	0.3	86	416.16	0.38	35.79
78	78	922	0.3	86	547.56	0.50	47.09
136	136	922	0.3	86	1664.64	1.53	143.16
157	157	922	0.3	86	2218.41	2.05	190.78
670	180	922	0.3	86	10854.00	10.01	933.44
200	200	922	0.3	86	3600.00	3.32	309.60
500	120	922	0.3	86	5400.00	4.98	464.40
68	68	922	0.3	86	416.16	0.38	35.79
68	68	922	0.3	86	416.16	0.38	35.79
170	170	922	0.3	86	2601.00	2.40	223.69
112	108	922	0.3	86	1088.64	1.00	93.62
.λ= SCALE SIZE OF THE ELECTRIC_VLSI LAYOUT PROGRAM							

Figure 8. Tabular method to design CMOS capacitances. This Excel sheet goes from X-Y dimensions to capacitance value.

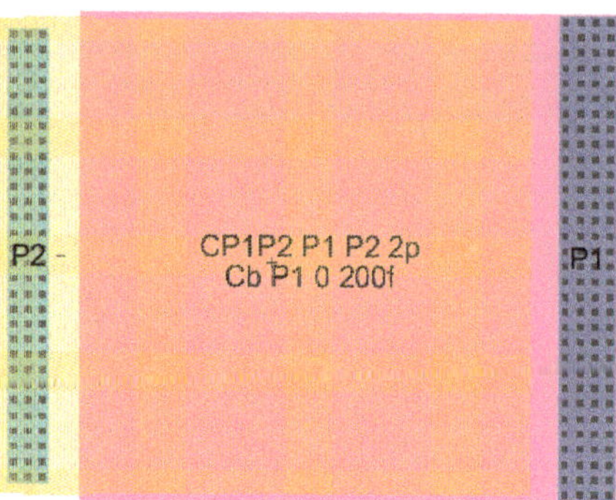

Figure 9. CMOS IC poly-poly 2 pF capacitance with squared layout from Electric-VLSI.

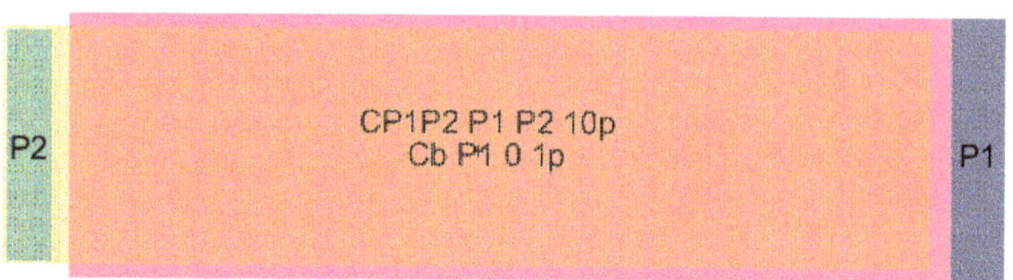

Figure 10. CMOS IC poly-poly 10 pF capacitance with rectangular X-Y layout from Electric-VLSI.

Process Specifications and Boundary Conditions

(Left margin, vertical: Design Process)

Differential Amplifier Step-by-Step Design Methodology
CMOS 600nm Process....Uses MOSIS data from C5_Models latest runs

	VT0	VDD	VSS	SR (V/μ)	Av (V/V)	CL (pF)	f.3dB	λ	minICMR	maxICMR	Pmax	Kp	Kn	
	0.8	2.5	-2.5	10	50	2	1E+05	0.04	-1	2	1	37.4	118	
step		Iss	max Iss	Min Iss	Pssmax	Rout max	Iss-sel	Vsg3	(W/L)3	(W/L)1	Vds5	(W/L)5	Vgs5	Vbias.5
1	Iss.for.SR	2.00E-05												
2	Max.Iss		2.00E-01		1.00E-04									
3	Min.Iss			3.14E-05		7.96E+05								
select	Iss						3.50E-05							
4	Ic_MAX							1.30	3.74					
select	(W/L)3								4.00					
5	Gain									1.19				
select	(W/L)1									1.00				
6	Ic_MIN										0.15	4.98		
select	(W/L)5											5.00		
7	Vbias												0.95	-1.55

Figure 11. Two-dimensional processing method to design a differential amplifier. From the initial cell (top-left), the conceptual design progress downward, step by step, and to the right to size each transistor.

Figure 12 shows the schematic from Electric-VLSI of a differential amplifier using results from the Excel two-dimensional method. In this case, every value sizing the MOSFETs means L/2 times or half the feature size of the technology. For instance, the middle twin transistors have a size of W = 120 × (0.5/2) = 30 μm by L = 2 × (0.5/2) = 0.5 μm. Furthermore, the SPICE code to perform DC and AC testing in this conceptual design of the device is shown in Figure 12. Once this schematic circuit model is tested, the next step is developing the layout. Differential amplifiers are the core of every instrumentation amplifier and their layout must be considered very carefully. Figure 13 shows the strategy recommended by J. Baker [4] to develop a common centroid layout.

With careful development by considering the DRC (direct rule checking) from Electric-VLSI CMOS 500 nm kit, the layout shown in Figure 14 is developed by using the evaluations and conceptual model obtained from the Excel two-dimensional method.

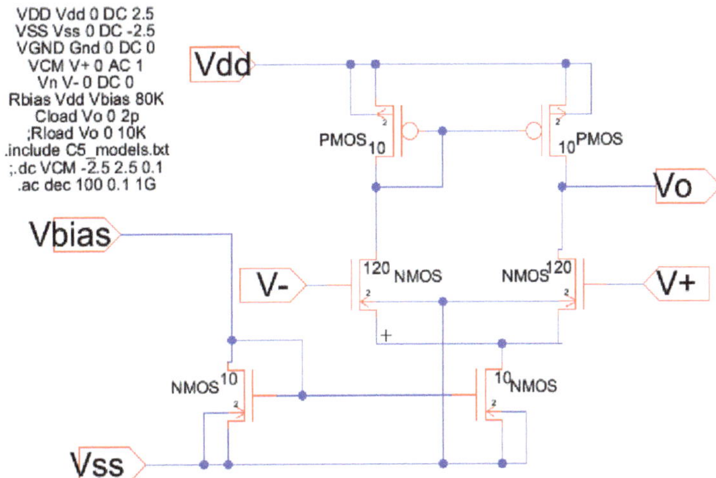

Figure 12. A differential amplifier schematic using CMOS 500 nm technology.

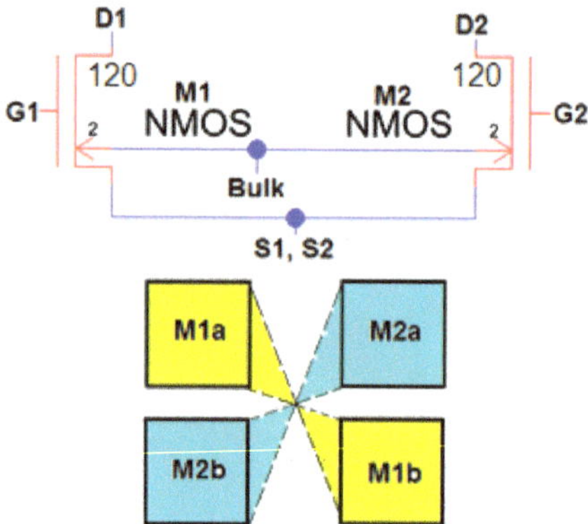

Figure 13. Common centroid layout recommended for big matched differential pair transistors [3,4].

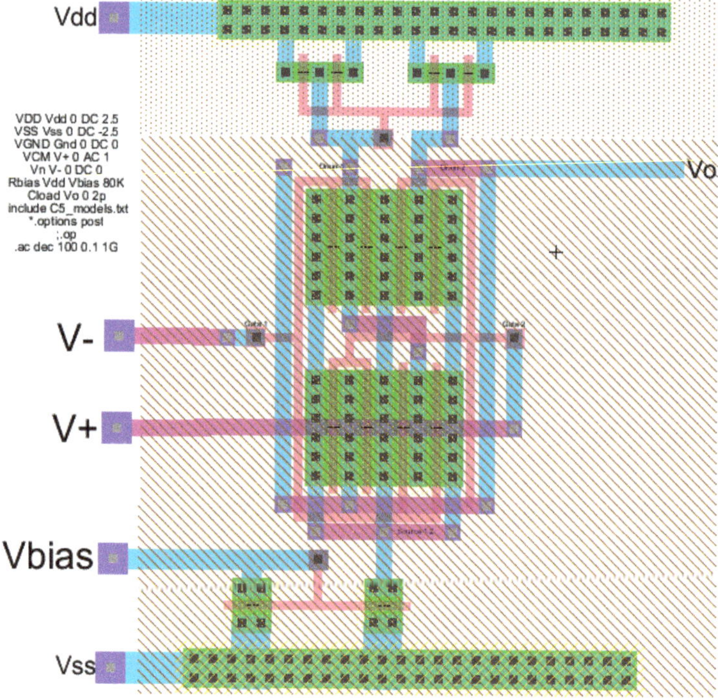

Figure 14. Conceptual design layout (using Electric-VLSI) of a differential amplifier from an Excel method.

5. Methodologies for Complete Amplifier Design

The Excel methodologies shown previously can be applied to develop conceptual designs of complete functional blocks such as a cascode amplifier with its bias circuit, and a two-stage operational transconductance amplifiers (OTA). The OTA amplifier includes a compensation capacitance which is necessary to ensure stability and reliable operation for the required frequency response. Even though the following design examples are not very specialized, the literature shows many examples of more specialized conceptual designs where microelectronic design has been extended [30–33]. The design flow that students must follow to perform the conceptual design of the microelectronic device is given as follows:

1. Select or create the basic structure. This step consists of obtaining the schematic showing the transistors and their connections. This diagram does not change through the design process unless the specs cannot be met.

2. Set up the Excel spreadsheet to show the specifications and boundary conditions in the first line. Make sure that the units are consistent all the way through the design process. Develop the decision process line by line using the equations and specs. The design flow goes downwards.

3. Select the DC currents and transistor sizes to meet specs. This is where the major design effort will go. Simulators are used to aid the designer in this phase. However, a rough performance of the circuit should be known a priori.

4. Physical implementation. Layout of the transistors, floor-planning the connections, pinouts, power supply buses, and grounds. Extraction of physical parasitic and re-simulation. Verify that the layout is a physical representation of the circuit.

5. Furthermore, in designing a multi-stage amplifier where two or three stages integrate the complete device, the following considerations are made:

    I. The characterization of the microelectronic device is the initial fundamental step in the analysis and design of the multi-stage amplifier.

    II. Ideal analysis provides the first insight to the circuit operation by performing circuit analysis over the external components.

    III. Practical models of the op-amp include static and dynamic parameters.

    IV. The Excel method helps in developing the overall step-by-step procedure which accounts for all the requirements and specifications

    V. The design of multi-stage devices involves the following boundary conditions and requirements:

    a. Boundary conditions: CMOS technology, process specs (V_{T0}, C_{ox}, K'), supply voltage and current range (V_{DD}, V_{SS}, I_{SS}), operating temperature (T_o) and range.

    b. Requirements: Gain (A_v, A_i), gain bandwidth (GB), settling time (T_s), slew rate (SR), input common mode range ($ICMR$), common mode rejection ratio ($CMRR$), power supply rejection ratio ($PSRR$), output voltage swing (v_{out} (max), v_{out} (min)), output resistance (R_{out}), offset voltage (V_{OS}), noise ($e_{out}{}^2$), layout area.

    c. Verify that intermediate stages couple correctly without causing instabilities such as the influence of mirror poles in the transfer function. The Excel method can be used to revise and iterate the process in search of a robust device.

    VI. Compensation techniques involve: Miller, feed-forward, and self-compensation schemes.

5.1. Cascode Amplifier with Bias Source

The cascode amplifier obtains a higher gain and output resistance than the traditional inverting amplifier stages. Typical design parameters are slew rate (SR), output swing, and power dissipation for a simple cascode stage. Figure 15 illustrates the Excel method to develop the conceptual design for a cascode amplifier. As mentioned before, the method illustrates the step-by-step evaluation and decision-making process downwards and the processing flow goes rightwards. Again, the CMOS Technology characteristics and specifications flow horizontally to the right and the CMOS design equations, from (5) to (10), flow downwards illustrating the step-by-step procedure. The evaluation of CMOS transistor size, W/L, goes along to fulfill the required specification.

Process Specifications and Boundary Conditions

Three Transistor Cascode Amplifier

CMOS 600nm Process....Uses MOSIS data from C5_Models latest runs

step	VT0	VDD	VSS	SR (V/µ)	Av (V/V)	CL (pF)	λ	Vo min	Vo max	Pmax	Kp	Kn	
	0.8	5	0	10	50	5	0.03	1.5	4	1	37.8	115.6	
		I-DS	max-ID	min-ID	Pssmax	ID-sel	(W/L)3	VGG3	(W/L)1	Vds1	Vds2	(W/L)2	VGG2
1	ID for SR	5.00E-05											
2	Max ID		2.00E-04		1.00E-03								
3	Min ID	5.00E-05		5.00E-05									
select	ID					5.00E-05							
4	Vo Max						2.65	3.26					
select	(W/L)3						3.00	3.2					
5	Gain								0.49				
select	(W/L)1								1.00				
6	Vo Min									0.93	0.57	2.66	2.27
select	(W/L)2											3.00	2.30

Figure 15. Two-dimensional processing method to design a CMOS cascode amplifier.

Figure 16 shows the schematic from Electric-VLSI of the conceptual design of the cascode amplifier which includes the bias network on the left of the stacked three MOSFET from the right. The bias circuit was implemented and evaluated in a separate analysis using the guidelines from J. Baker [4]. Once this schematic circuit model is tested, the next step is to develop the layout.

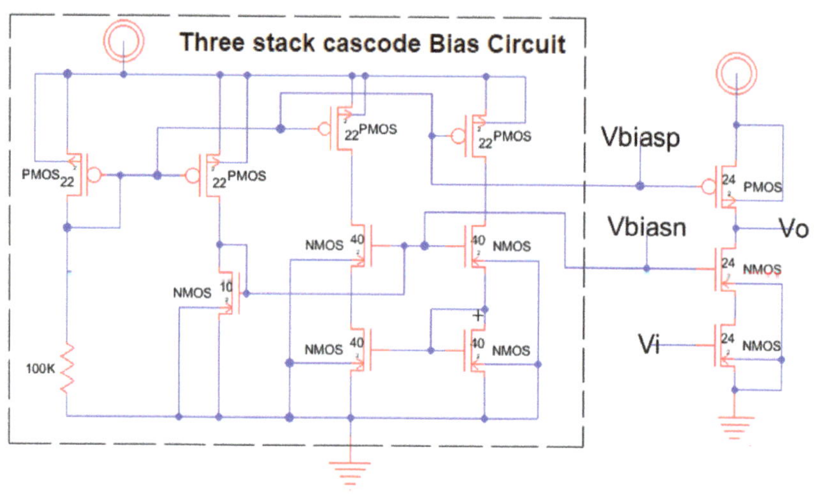

Figure 16. A three stack cascode amplifier schematic using CMOS 500 nm technology.

The layout is developed using Electric-VLSI and includes both the bias circuit and the three stacked transistors as shown previously in Figure 16. Figure 17 illustrates the layout design with the corresponding V_{dd} = 5 V and Vss = 0 (ground). This layout shows the transistors in horizontal layouts where the lower two levels include N-type transistors, and the upper level includes the P-type transistors which appear in the circuit schematic from Figure 16. The layout also shows the p-well and n-well over the lower two N-channel transistor levels and the upper P-channel transistor level, respectively.

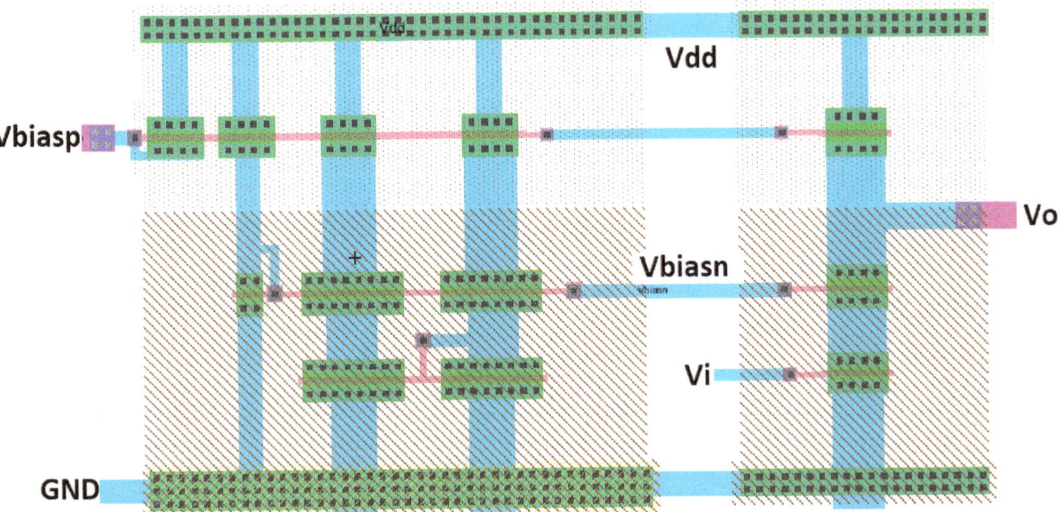

Figure 17. Conceptual design layout (using Electric-VLSI) of a three stage cascode from an Excel method.

5.2. OTA with Miller Compensation

The development of a conceptual model for an operational transconductance amplifier with Miller compensation has an additional complexity of calculating the feedback capacitance that operates the amplifier in a stable and reliable regime. The phase margin PM > 50° and the sizing of more than 10 transistors make the Excel method larger. Figure 18 shows the Excel spreadsheet workout with the evaluation of transistor sizes and dominant pole calculations.

Again, the procedure shows the conceptual design strategy mentioned before, the CMOS technology characteristics and specifications flow horizontally to the right and the CMOS design equations, from (5) to (10), flows downwards illustrating the step-by-by step procedure. In this case, Figure 18 has two downward decision flows. The first one evaluates the design with the sole calculation of the Miller compensation capacitance to provide the required phase margin and stability criteria [3,4,17,27]. The second design decision flow determines the size of a transistor to locate a right-hand side pole (RHP) exactly to cancel the second pole. This way the dominant pole will be extremely alone well inside the gain bandwidth and the amplifier will have, even higher, phase margin PM. Figure 19 shows the Electric-VLSI layout schematic of the conceptual design obtained for the 500 nm CMOS technology. Figure 20 illustrates the layout of the OTA amplifier with the area dominance of the compensation capacitance $C_c = 3$ pF. Those capacitances are extremely large and, in this case, the layout generated has a squared shape, like the ones developed with a single straight Excel methodology for capacitors. The layout shows also the common centroid differential amplifier stage developed before and the big, 125 μm/1 μm, p-channel MOSFET transistor right above the output voltage of the amplifier.

Process Specifications and Boundary Conditions

1 · Design Flow · 2

Cox	VT0	VDD	VSS	SR (V/µ)	Av (V/V)	CL (pF)	GB	λ p	λ n	minICMR	maxICMR	Pmax	Kp	Kn	Vo min	Vo max	PM
DESIGN OF A TWO STAGE OTA USING L=0.6 MICROMETER. THE FIRST TWO ROWS ARE SPECIFICATIONS																	
0.25	0.8	2.5	-2.5	10	3000	10	5.00E+06	0.03	0.03	-1.6	2	1	37.8	115.6	-2	2	60
step		Cc	I5	(W/L)3	p3-rad	p3-Hz	10*GB	gm1	(W/L)1	Vds5	(W/L)5	gm6	gm4	(W/L)6	I-6	Pdiss	(W/L)7
1	min Cc	2.20														mW	
select		3															
2	I5		3E-05														
select			3E-05														
3	maxICMR			19.84													
select				20													
4	check M p				2.25E+10	3.58E+09	5.00E+07										
check	p3>10GP					p3-Hz>>	10*GB										
5	gm1	(W/L1)						9.42E-05	2.56								
select									3								
6	Vds5	(W/L)5								-0.344	4.38						
select											5						
7	gm6	(PM)	(W/L)6									9.42E-04	1.51E-04	125.10			
select														125			
8	I6	Pdiss													9E-05	0.62	
check																OK	
9	(W/L)7																15.65
select																	16
10	check Vo min									Av DB			Va min		-2.18		
	check Av					Av		17500		84.861							
RHP-ZERO COMPENSATION USING AN ADDITIONAL TRANSISTORS: M8, M9, M10 AND M11																	
11	I9, I10, I11			select current		3.00E-05			3.00E-05	(W/L)11 =		39.94					
select	(W/L)11											40					
12	(W/L)9					(W/L)10=	39.94			(W/L)9 =		5					
select							40					5					
13	(W/L)8									(W/L)8 =		28.87					
select												30					
14	check		VSG10	1.00	RZ	4600.1415	z1	-9.42E+07	-1.499E+07				p2	-94200000	OK for RHP-ZERO COMP		

Figure 18. Conceptual OTA design with two downward decision flows.

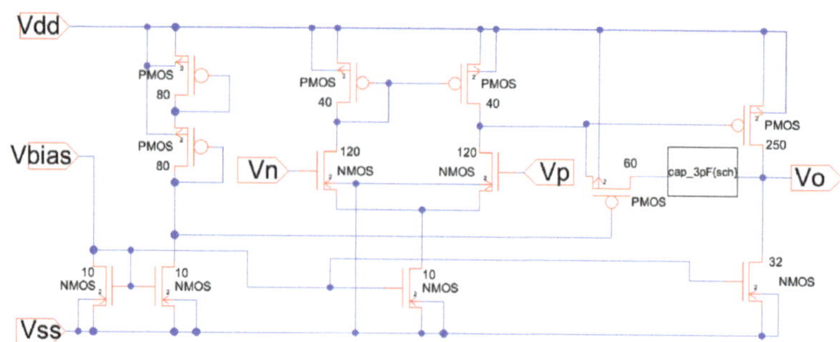

Figure 19. Electric-VLSI schematic of the CMOS OTA with RHP-zero compensation.

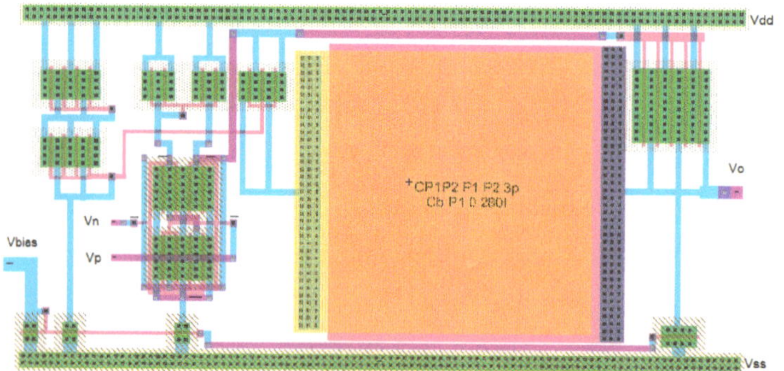

Figure 20. Electric-VLSI layout of the CMOS OTA with RHP-zero compensation.

5.3. Performance Simulation Tests

The Excel methods are used to synthesize the conceptual designs of integrated circuits and devices to teach and develop successful strategies that can be repeated for different CMOS technologies. This section will compare the expected specifications defined initially with the schematic and layout results from SPICE simulations run by Electric-VLSI using the LTSpice program as a kernel.

Four conceptual design cases are analyzed which are part of a formal course in microelectronics [28]. The design problems are:

1. Differential amplifier using common centroid layout.
2. The three-stack MOSFET cascode amplifier using the corresponding bias power reference.
3. The two-stage OTA stage with Miller compensation and having the RHP cancel the second dominant pole.
4. A three-stage op-amp using shunt feedback output stage to enhance the output resistance.

The differential amplifier results are summarized in Table 5. This amplifier stage is shown in Figures 11 and 13 and resolved using the two-dimensional processing method from the Excel methodology illustrated in Figure 10. The results comply with all the specifications established for the conceptual model using the CMOS 500 nm technology as illustrated in Table 5. Some parameters differ slightly because of the iterative nature of the design decision flow in which the number of requirements is higher than the number of degrees of freedom available: three transistor sizes (W/L) and the operating Q point of the current-sinking at the lower transistor in Figure 13.

Table 5. Specifications and measured values from simulation tests in the CMOS Differential amplifier.

Specification	Requirement	Simulation Measured Value
CMOS process	0.5 μm	0.5 μm
Supply voltage	5 V rail to rail	$\pm$2.5 V
Supply current	>30 μA	49.2 μA
Gain	>30 dB	33.7 dB
Offset voltage	<20 mV	19.57 mV
f_{-3dB}	>100 KH ($C_L = 2$ pF)	812.8 KH
Slew Rate	>10 V/μs ($C_L = 2$ pF)	-25 V/μs and $+27$ V/μs
Power Dissipation	<1 mW	0.492 mW

Note: Differences are between accepted tolerances of +/−10% for the conceptual model development.

Table 6 illustrates the results for the cascode amplifier conceptual model developed using the Excel method shown in Figure 14. Again, this amplifier stage, shown in Figures 15 and 16, describes the fulfillment of all the specifications established for the conceptual model using the CMOS 500 nm technology. The power dissipation shown for this amplifier includes the three stack of transistors and the bias reference network shown in Figures 15 and 16. Again the number of requirements is higher than the number of degrees of freedom available: three transistor sizes (W/L) and the operating Q point of the current-sinking at the lower transistor in Figure 16. The two previous amplifiers, differential, and cascode are not used as independent amplifiers, but they are part of a larger multistage amplifier or microelectronics functional block. Therefore, the conceptual models for a highly specialized microelectronic device contain 2, 3, or more of those primitive amplifiers described previously. Now we will present results for the conceptual model of a two-stage OTA and of a three-stage operational amplifier. Table 7 shows results for the conceptual model of a two-stage OTA that includes a differential amplifier and an inverting amplifier that drives the load. This was the amplifier illustrated by Figures 18 and 19 and developed using the methodology of Figure 17. This OTA conceptual design includes additional specifications such as offset voltage, output swing, phase margin, power supply rejection ratio (PSRR), gain-bandwidth GB, and settling time. This is a preview of the project that the students in the microelectronics course develop at the end of the semester.

Table 6. Specifications and measured values from simulation tests in the CMOS cascode amplifier using using 0.5 μ porocess.

Specification	Requirement	Simulation Measured Value
Supply voltage	5 V rail to rail	$V_{dd} = +5$ V and $V_{ss} = 0$ V
Supply current	>30 μA	52 μA
Gain	>26 dB	31.1 dB
f$_{-3dB}$	>200 KH ($C_L = 5$ pF)	361 KH
Slew Rate	>10 V/μs ($C_L = 5$ pF)	-300 V/μs and $+$ 10.45 V/μs
Output swing	from 1.5 to 4.0 V	from 0.66 to 4.03 V
Power Dissipation	<1 mW	0.996 mW

Note: Differences are between accepted tolerances of $+/-10\%$ for the conceptual model development.

Table 7. Specifications and measured values from simulation tests in the CMOS 2-stage OTA using 0.5 μ process.

Specification	Requirement	Simulation Measured Value
Supply voltage	5 V rail to rail	±2.5 V
Supply current	>30 μA	30 μA
Gain	>80 dB	80 dB
Gain-Bandwidth	>10 MH	14.7 MH
Slew Rate	>10 V/μs ($C_L = 2$ pF)	-10 V/μs and $+$ 12 V/μs
ICMR	> ±1.5 V	-2.3 to $+$ 2.4 V
Offset voltage	< ±2 mV	-0.164 mV
PSRR	>70 dB	80 dB
Output Swing	> ±2	-2.25 V and $+$ 2.25 V
Phase Margin	>70°	83°
Power Dissipation	<1 mW	0.896 mW

Note: Power dissipation includes the bias current for both stages.

Finally, Table 8 illustrates the results obtained from a conceptual design of a 3-stage op-amp using an additional third stage with a shunt feedback scheme to reduce the output resistance of the device. The results are good with a low value with the negative slew-rate (SR) of -9 V/μs which needs to be improved from this conceptual design developed using the methodologies discussed in this paper.

Table 8. Specifications and measured values from simulation tests in the CMOS 3-stage op-amp using 0.5 μ process.

Specification	Requirement	Simulation Measured Value
Supply voltage	5 V rail to rail	±2.5 V
Supply current	>30 μA	73 μA
Gain	>90 dB	80.19 dB
Gain-Bandwidth	>10 MH	24.9 MH
Slew Rate	>10 V/μs ($C_L = 2$ pF)	-9V/μs and $+$ 12.45 V/μs
ICMR	> ±1.5 V	$-$ 2.2 to $+$ 1.84 V
Offset voltage	< ±2 mV	0.036 mV
PSRR	>70 dB	80 dB
Output Swing	> ±2	-2.25 V and $+$ 2.25 V
Phase Margin	>70°	126.4°
Power Dissipation	<1 mW	1.095 mW

Note: Power dissipation includes the bias current for the three stages.

The schematic of the three-stage op-amp conceptual design is shown in Figure 21. This figure illustrates the main sub-components of the device: bias and reference voltage circuit, first stage differential amplifier, Miller and right-hand plane zero (RHP) compensation for maximum stability, inverting amplifier second stage, and the output push/pull with shunt feedback differential amplifier that provides a lower output resistance in the device.

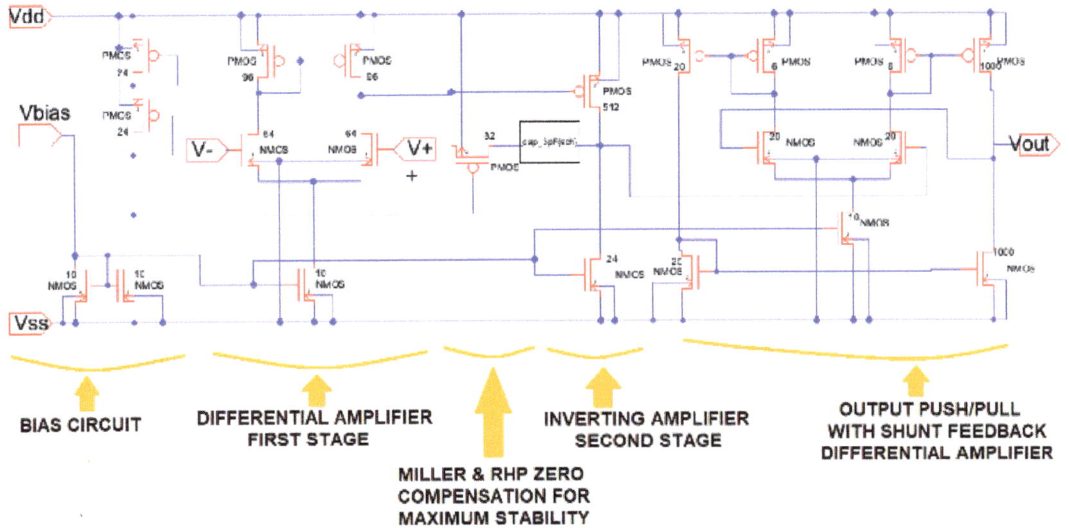

Figure 21. Electric-VLSI schematic of the 3-stage CMOS op-amp with RHP-zero compensation.

The conceptual design developed in this paper shows one of the major steps in designing analog integrated circuits (IC) for electronic instrumentation devices required by electronics, biomedical, robotics, and computer engineering majors. In analog IC design, a good combination of function or application with IC technology is necessary to obtain a successful solution. The Excel methodologies are an additional tool to validate and verify the conceptual model required before the device is sent to the foundry facility. Analog IC design consists of three major steps [3]: electric design, physical design (layout), and test design (testing). Engineers and designers must be flexible, use techniques such as the Excel methods, and have a skill set that allows them to simplify and understand a complex conceptual design problem. In microelectronics, device IC design is driven by improving technologies rather than new technologies [3]. The engineer should be aware that sometimes analog systems applications, where speed, area, or power, have certain advantages over the digital approach. Even using Excel methods, deep-submicron (DSM) technologies offer great challenges to the creativity of engineers and designers of IC microelectronic devices.

6. Analysis of Results

To further increase the potential of using Excel methods in developing instrumentation amplifiers for biomedical applications, we examined a project case study using the methodology to design a 3-stage amplifier having a low voltage and low power operation in the strong inversion zone. Undergraduate students during the spring semester of 2021 in the microelectronics course at Tecnológico de Monterrey developed a project where they used the methods learned in class [34]. This design was going to be used as a subsystem of a bioinstrumentation amplifier required in sensor signal conditioning applications.

The project consisted of the development of a three-element instrumentation amplifier for biomedical applications. Figure 22 illustrates the basic scheme where two op-amps receive the differential mode input signals, and a third op-amp changes the signal from differential mode to single-ended mode referenced to ground.

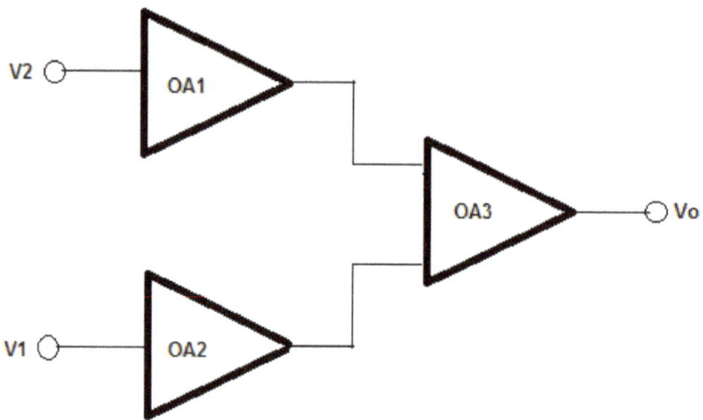

Figure 22. Classic topology for the instrumentation amplifier used in biomedical applications.

In Figure 22, each op-amp (OA1, OA2, and OA3) must be selected from possible topologies seen in class to achieve certain performance characteristics. Those op-amps have the following components:

1. First stage differential amplifier;
2. Second stage common source amplifier;
3. Third stage push-pull output stage;
4. Power source to provide the required bias currents and voltages.

Figure 23 shows a block diagram of each op-amp with all the functional parts of the system. For the power source, the students have the option of developing a high-performance bandgap reference source that is stable with respect to variations in supply voltage, temperature, and noise.

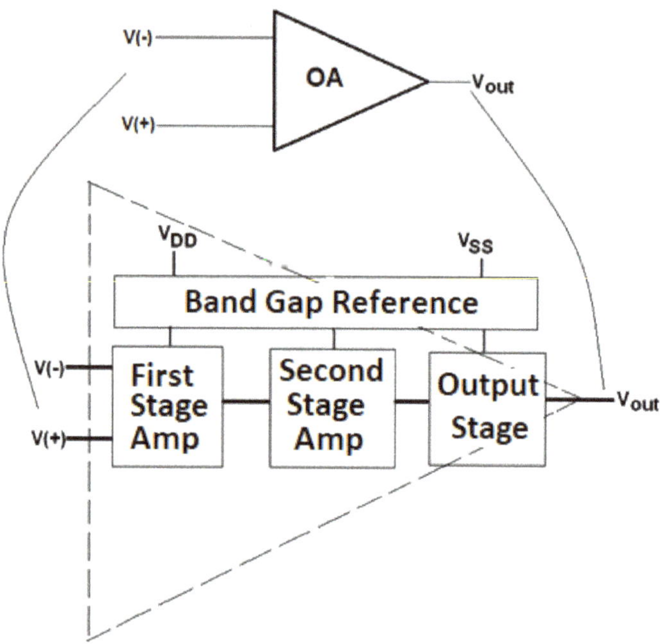

Figure 23. Overall block diagram of each operational amplifier.

The requirements and specifications provided in Table 9 are to be met for each of the op-amp blocks.

Table 9. Proposed specifications for the op-amp design for biomedical applications.

Specification	Requirement	Comment
CMOS Process	0.5 μm	
Supply Voltage	From 2.0 V to 5.0 V rail to rail	+/−1.0 V to +/−2.5 V (for Vdd/Vss)
Supply Current	<100 μA	
Temperature range	0 to 70 °C	
Gain	>90 dB	
Gain Bandwidth	>10 MH	
Settling time	<0.5 μs	
Slew rate	>10 V/ μs	Average of SR+ and SR-
ICMR	>0.8 V to >+/−1.5 V	Depends upon the supply rails
PSRR	>70 dB	
Output swing	>+/−2 V	
Output resistance	<100 Ohms	
Offset Voltage	From +/−0.8 V to +/−2 mV	Depends upon the supply rails
Noise	<50 nV/sqrtH at 100 H	
PM	>70°	
Power dissipation	<1 mW	

The students began with a literature survey before selecting the device to develop from the conceptual requirements, theoretical development, schematic development, and layout implementation using Electric_VLSI [35–38]. Instead of adding a third stage at the output of the two-stage OTA device, they added a second stage differential-amplifier with compensation between the first differential stage and the output stage. This results in improving the frequency response and sacrificing some gain and having a better phase margin for even robust design. Even with this modification, the simulations show an increase in the overall gain compared to the last three stage designs (Section 5). The compensation capacitor Cc and Rz are kept for both differential stages, and the output terminals of both compensation elements are connected all the way to the output of the circuit. The methodology generates the transistor dimensions and capacitor values illustrated in Tables 10 and 11.

Table 10. Transistor dimensions for the new three-stage conceptual design.

Transistor Dimensions (W/L) Considering L = 500 nm	
M1, M2	8
M3, M4	7
M5	5
M6	130
M7	22.5
M8	27
M9, M12	5
M10, M11	15

Table 11. Capacitors and resistors for the new three-stage conceptual design.

Cs are in pF and Rs are in KW	
C_L	2
C_C	0.5
R_{Bias}	265
R_z (Implemented by M8)	5.22

Considering the transistor dimensions for this new three stage amplifier design, the first cut expected parameters are calculated as follows, assuming long channel models.

1. Slew rate.

$$SR = \frac{I_5}{C_c} = \frac{10\mu}{0.5p} = 20 \text{ V}/\mu s \tag{17}$$

2. Estimated amplifier gain.

$$A_{v0} = \left(\frac{g_{m2}}{g_{m2} + g_{m4}}\right)^2 \cdot \frac{g_{m6}}{g_{ds6} + g_{ds7}} = \left(\frac{g_{m2}}{I_5/2(\lambda_2 + \lambda_4)}\right)^2 \cdot \frac{g_{m6}}{I_6(\lambda_6 + \lambda_7)} \tag{18}$$

$$g_{m2} = \sqrt{2\beta_2 I_2} = \sqrt{2 \times 115.6\mu \times 8 \times 5\mu} = 96.17 \ \mu S \tag{19}$$

$$g_{m6} = 942.5 \ \mu S \tag{20}$$

$$A_{v0} = \left(\frac{96.16\mu}{5\mu(0.04 + 0.05)}\right)^2 \cdot \frac{942.5\mu}{92.14\mu(0.04 + 0.05)} = 134.3 \ \text{dB} \tag{21}$$

3. Power dissipation.

$$P_{diss} = (2I_5 + I_7)(V_{dd} + |V_{ss}|) = (2 \times 10\mu + 92.14\mu)(2 \times 1.8) = 403.7 \ \mu W \tag{22}$$

4. Output swing.

$$V_{out}(min) = V_{ss} - V_{DS7}(sat) = -1.8 + \sqrt{\frac{2I_7}{k_N(W/L)_7}} = 1.8 + \sqrt{\frac{2 \times 92.14\mu}{115.6\mu \times 22.5}} = -1.53 \ \text{V} \tag{23}$$

$$V_{out}(max) = V_{dd} - V_{SD6}(sat) = 1.8 - \sqrt{\frac{2I_6}{k_P(W/L)_6}} = \sqrt{\frac{2 \times 92.14\mu}{37.8\mu \times 130}} = 1.607 \tag{24}$$

5. Input common mode range, ICMR.

$$(W/L)_5 = \frac{2I_5}{k_N V_{DS5}(sat)^2} = \frac{2 \times 10\mu}{k_N\left(-V_{ss} + ICMR^- - \sqrt{\frac{I_5}{\beta_1}} - V_{T1}(max)\right)^2} \tag{25}$$

$$ICMR^- = -0.93V \tag{26}$$

$$(W/L)_3 = \frac{I_5}{k_P(V_{dd} + ICMR^+ - |V_{T3}|(max) + V_{T1}(min))^2} \tag{27}$$

$$ICMR^+ = 1.69V \tag{28}$$

6. Output resistance.

$$R_{out} = \frac{1}{g_{ds6} + g_{ds7}} = \frac{1}{I_6(\lambda_N + \lambda_P)} = \frac{1}{92.14\mu(0.04 + 0.05)} = 120.5 \ \text{K}\Omega \tag{29}$$

Afterward, the circuit schematic and layout are developed in Electric VLSI to perform simulations tests. The compensation capacitor is designed as a poly-poly2 capacitor. The area for the capacitor is 78 by 78 lambda, with a 47 fF parasitic capacitance. Figure 24 shows the Excel first cut approximation used for the first part of the conceptual design.

DESIGN OF A LOW VOLTAGE TWO STAGE OTA USING L=0.6 MICROMETER. THE FIRST TWO ROWS ARE SPECIFICATIONS

Cox	VT0	VDD	VSS	SR (V/µ)	Av (V/V)	CL (pF)	GB	λ_p	λ_n	minICMR	maxICMR	Pmax	Kp	Kn	Vo min	Vo max	PM
0.25	0.8	1.8	-1.8	20	3000	2	2.90E+07	0.05	0.04	-0.7	1.7	1	37.8	115.6	-1.6	1.7	60
step		Cc	I5	(W/L)3	p3-rad	p3-Hz	10*GB	gm1	(W/L)1	Vds5	(W/L)5	gm6	gm4	(W/L)6	I6	Pdiss	(W/L)7
1	min-Cc	0.44		(W/L)4												mW	
select		0.5															
2	I5		0.00001														
select			0.00001														
3	maxICMR			6.613757													
select				7													
4	check M-p				2.19E+10	3.49E+09	2.90E+08										
check	p3>10GP					p3-Hz>>	10*GB										
5	gm1	(W/L)1						9.11E-05	7.17								
select									8								
6	Vds5	(W/L)5								0.19601	4.50						
select											5						
7	gm6	(PM)	(W/L)6									9.11E-04	5.14E-05	123.92			
select														130			
8	I6	Pdiss													8E-05	0.3397	
check																OK	
9	(W/L)7																37.992
select																	35
10	check Vo min										Av DB		Vo min		-1.60		
	check Av						Av	24267			87.7002						
RHP-ZERO COMPENSATION USING AN ADDITIONAL TRANSISTORS: M8, M9, M10 AND M11																	
11	I9, I10, I11			select current		1.00E-05		1.00E-05				(W/L)11 =	15.41				
select	(W/L)11												15				
12	(W/L)9					(W/L)10=	15.41					(W/L)9 =	5				
select							15						5				
13	(W/L)8											(W/L)8 =	25.65				
select													27				
14	check		VSG10	0.99	RZ	5490.885	z1	-4.55E+08	-7.246E+07			p2	-5E+08	OK for RHP ZERO COMP			

Figure 24. Excel methodology for the first part two-stage OTA conceptual design, before connecting the third stage to achieve higher gain and robust conceptual design.

The Excel design spreadsheet shown in Figure 24 replicates the strategy which has been proposed in this article. The CMOS technology characteristics and specifications flow horizontally to the right and the CMOS design equations, from (5) to (10), Tables 1 and 2, flow downwards illustrating the step-by-by step procedure. Figures 25 and 26 illustrate the schematic and layout diagrams for the three-stage amplifier designed using the methodology shown and initiating with the conceptual design equations coming from the long channel model.

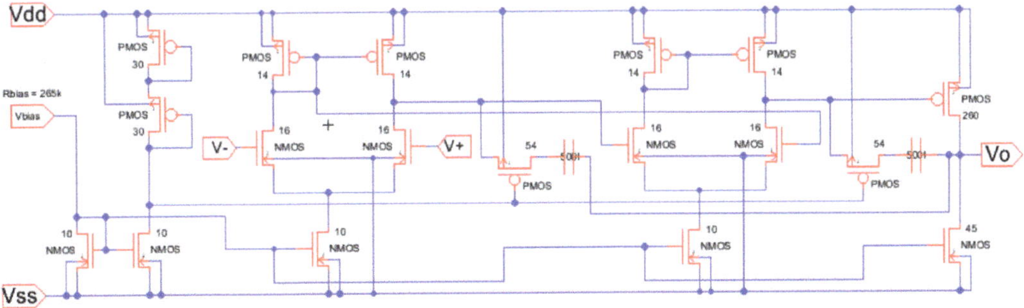

Figure 25. Schematic diagram from Electric_VLSI for the three-stage low-power high-gain operational amplifier.

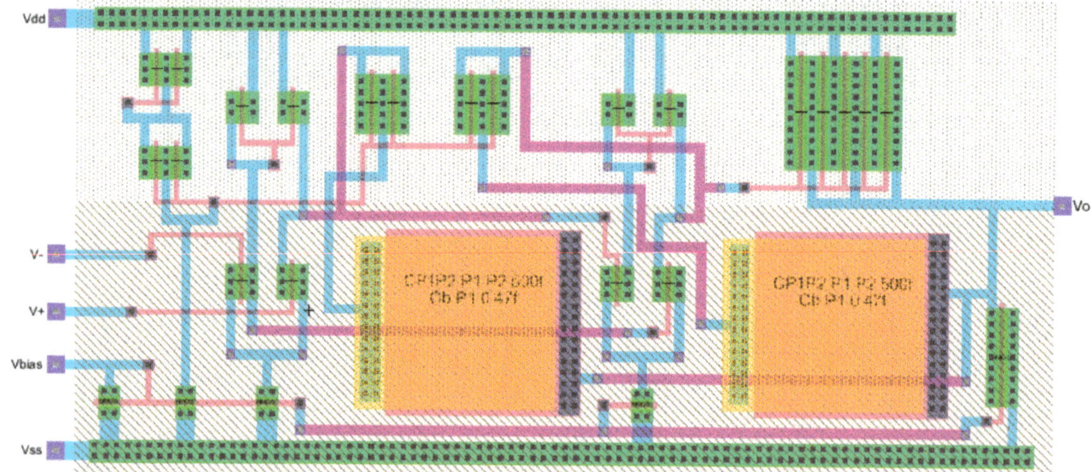

Figure 26. Layout diagram from Electric_VLSI for the three-stage low-power high-gain operational amplifier.

The output simulation runs show results that comply with the major specifications required by the design. Figure 27 illustrates the layout frequency sweep where the dB gain, gain bandwidth, and the phase margin are displayed. The DC gain obtained is 91.7 dB, the gain bandwidth GB is 46 MH and the phase margin is now close to 94°, which makes a very robust device.

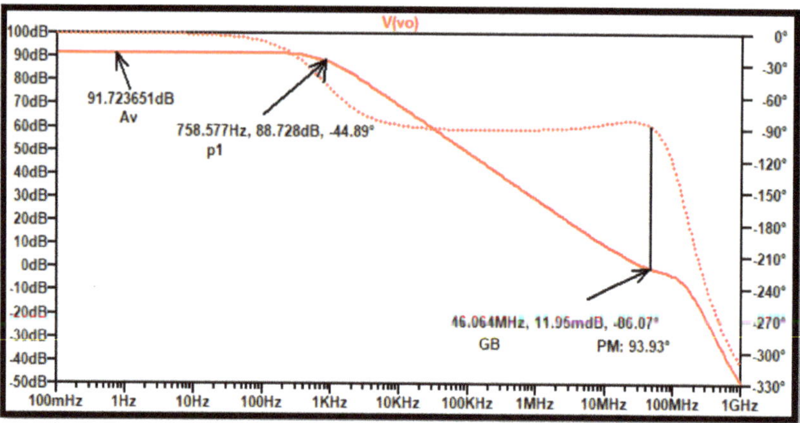

Figure 27. Frequency sweep using. AC analysis in SPICE for the three-stage low-power high-gain op-amp Layout.

Figure 28 shows the time domain simulation to verify the slew rate response. The graph shows $SR^+ = 21.47$ V/µs and $SR^- = -19.35$ V/µs which is much more than the required response.

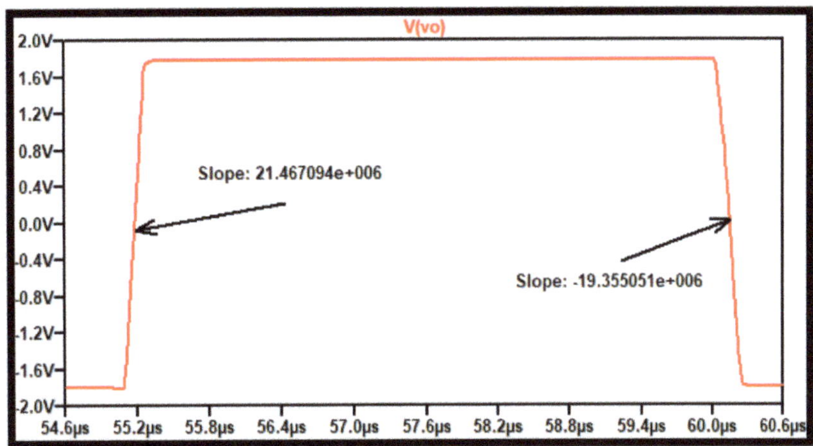

Figure 28. Time domain response using transient analysis in SPICE for the three-stage low-power high-gain op-amp layout.

To summarize the complete results for all the theoretical and expected parameters, Table 12 compares:

1. Desired requirements for the conceptual design;
2. Theoretical calculations using the Excel method;
3. Simulation results of the schematic circuit;
4. Simulation results of the layout circuit.

Table 12. Comparison of specification, theoretical, schematic and layout results for the three-stage, low-power high-gain operational amplifier conceptual design.

Specification	Requirement	Theory Calculation	Schematic Simulation	Layout Simulation
I_{SS}	10 µA	10 µA	9.72 µA	9.72 µA
A_{v0}	>90 dB	134.3 dB	91.84 dB	91.72 dB
GB	>30 MH	40 M	50.87 M	46.1 M
SR_+	>20 V/µ	*	21.7 V/µ	21.1 V/µ
SR_-	>|20 V/µ|	*	19.6 V/µ	19.4 V/µ
PM	> 70°	*	109.5°	93.9°
P_{diss}	< 1 mW	0.403 mW	0.675 mW	0.661 mW
ICMR-	−0.8 V	−0.93 V	−0.816 V	−0.81 V
ICMR+	1.4 V	1.69 V	1.632 V	1.62 V
$V_{out}(max)$	1.4 V	1.61 V	1.35 V	1.34 V
$V_{out}(min)$	−1.4 V	−1.53 V	−1.35 V	−1.34 V
PSRR-	>70 dB	*	88 dB	88.1 dB
PSRR+	>70 dB	*	97 dB	97.9 dB
CMRR	>90 dB	*	128.8 dB	123.4 dB
Noise	<0 nV/$\sqrt{H}$	*	20.9 nV/$\sqrt{H}$	20.9 nV/$\sqrt{H}$

(*) no anticipated calculations were made for those parameters.

The results compare well to the specified requirements except for the maximum positive swing of the output signal which does not reach the specified value of 1.4 V. Furthermore, the power dissipation is slightly higher than anticipated by the theoretical calculations of 0.403 mW. However, very good results are obtained in gain bandwidth (GB), CMRR, PSRR, phase margin PM, and noise. The requirements for slew rate (SR) and DC gain (A_{v0}) are barely achieved. Some of those specifications could have been obtained with additional calibration and refinements in the final conceptual design.

The student team [34] decided to compare the conceptual design of this low-power operational amplifier with one that appears in the literature with similar characteristics and is used in similar biomedical applications [35]. Table 13 shows this comparison where the supply voltage and the power consumption appear higher in our design. The load capacitance was a specification given by the instructor. Slew rate and bandwidth show a significantly higher performance in our design; however, those were parameters also required by the instructor. The chip area is twice in our design, however, in this case, the instructor did not have any restrictions here and no optimization was performed to reduce the layout artwork whatsoever. Table 13 results, however, illustrate comparisons under different specifications such as the load capacitance and voltage supply. Reference [35] used $C_L = 30$ pF and ±1.65 V, whereas our example used $C_L = 2$ pF and ±1.8 V, respectively. Some of those conditions were imposed by the course instructor. These changes produced large differences in closed-loop bandwidth and slew rate (SR) as seen by Table 11 results. Furthermore, the power dissipation and chip area obtained here are twice the values obtained by reference [35]. Another way to reduce the power dissipation and voltage supply could have been if they have designed the amplifier to operate in the subthreshold zone. However, the model equations in the Excel method would need to be changed to include the subthreshold model whatsoever. This comparison is an excellent way to encourage confidence in students about the design of integrated circuits at the nanoscale level [36].

Table 13. Comparison of the three-stage op-amp conceptual design with Lópes-Martin´s design [35] which appears in the technical literature using CMOS 500 nm technology.

Parameter	Lópes-Martin et al. [35]	This Study
CMOS Technology μm	0.5	0.5
Supply V	±1.65	±1.8
Power consumption mW	0.26	0.67
Load capacitance pF	30	2
SR V/μs	2	20.19
Silicon area mm^2	0.02	0.04
Gain bandwidth MH	1	46.06

A system simulation test was performed by students with an instrumentation amplifier working with a gain of 60 dB or 1000 *v/v* using the traditional three op-amp topology. Figure 29 shows the Icon-View of the simulation experiment having three op-amps (3S_OA) and the required resistors to provide the gain factor and the bias of the device. The figure shows on the left-hand side the SPICE code to run the frequency domain test (.ac dec 100 0.1 1 G). Furthermore, the MOSFET models command < include C5_models.txt> describes the 500 nm technology used in this case.

The frequency-domain tests show a DC gain very close to the required for this instrumentation amplifier. Figure 30 shows the output voltage graph with a measured gain of 59.7 dB, approximately 966 V/V, which shows less than 5 % error with respect to the required 1000 V/V gain. This gain goes along to up to 10 KH in bandwidth.

Finally, a time-domain system test for the instrumentation amplifier was developed to find the maximum output symmetrical swing for the device when amplifying the biomedical sensor signal. This transient test was performed with a 1.8 mV amplitude at 1000 H differential mode signal (V2-V1) at the input. Figure 31 shows the output signal with a+1.7323 V to −1.735 V swing which also shows the gain of 966 *v/v* for the instrumentation amplifier configuration.

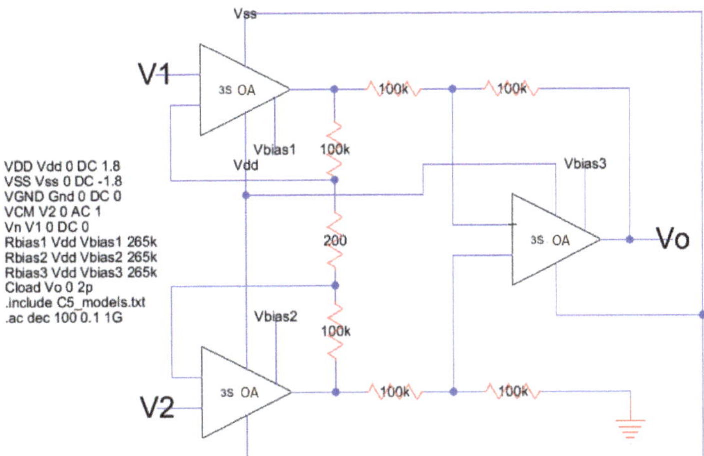

Figure 29. Instrumentation amplifier system simulation test using icon-view with Electric_VLSI.

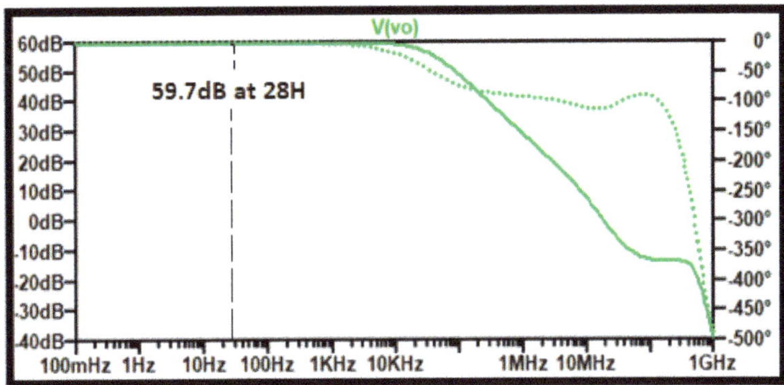

Figure 30. SPICE AC frequency domain test for the instrumentation amplifier.

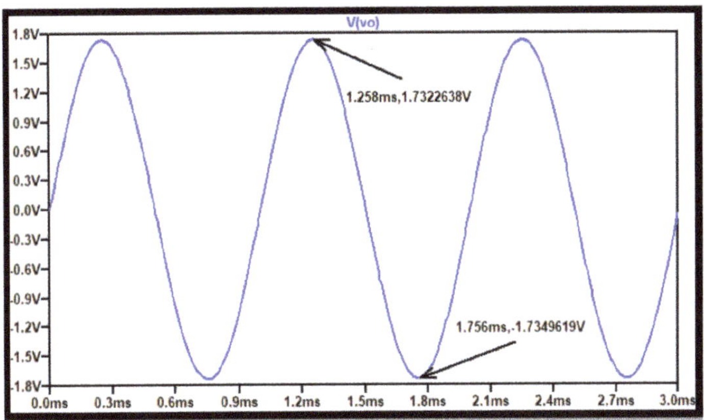

Figure 31. Time domain test for the instrumentation amplifier, where the maximum rail to rail swing is shown.

The designed operational amplifier and the instrumentational amplifier have good features for biomedical portable sensor conditioning applications. ECG and EEG biopotentials can be processed, and a good front device can be implemented using the conceptual design generated using these methods. The design of micro-power-operational amplifiers to fulfill biomedical instrumentation characteristics has become a huge yet complicated research task with great opportunity areas for improvement and innovation. The design proposed by the students offers further improvement opportunities (like reducing output resistance) yet possesses interesting features that would hopefully serve for its application in sensor signal conditioning in biomedical instrumentation.

Another application of the low power operational amplifier is to perform signal conditioning in signals coming from CMOS-MEMS sensors [39–42] where the CMOS amplifiers are used as the standard electronic system maneuvering platform. For instance, in [42] the low power operational amplifier is used as a high input impedance instrumentation amplifier to condition signals coming from a micro-hotplate in either Pirani, Temperature, or Gas CMOS-compatible MEMS sensors. Three operational amplifiers, designed in this paper, are set up as instrumentation amplifiers like the one shown in Figure 29. This configuration is ideal for this multifunctional sensor platform application, particularly for the results obtained by the temperature sensor as shown in Figure 32. Figure 33 shows the new instrumentation amplifier configuration used to condition the signal coming from the multiplatform temperature sensor [42] to obtain the results shown in Table 14. Figure 34 illustrates the linear conditioning performed by the instrumentation amplifier designed using the low power OTA developed by the students.

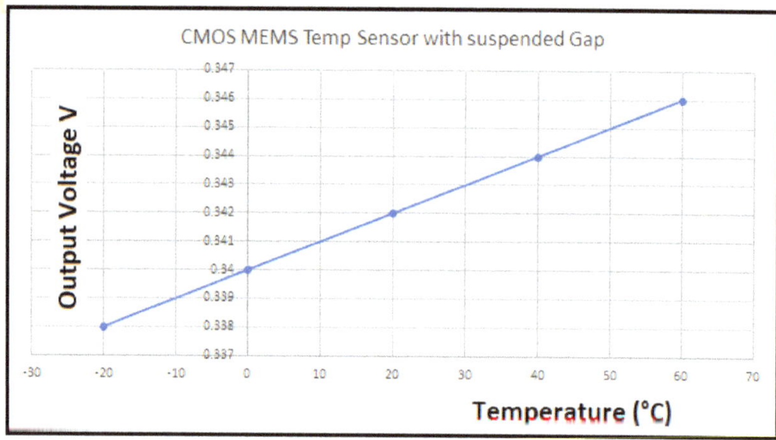

Figure 32. Output voltage for a CMOS MEMS temperature sensor with suspended gap from reference [42].

Table 14. CMOS-MEMS signal conditioning for temperature sensor.

Temperature (°C)	V2 (V)	V1(V)	V2-V1 (V)	Gain (V/V)	Vref (V)	Vo (V)
−20	0.338	0.34	−0.002	100	0	−0.2
0	0.34	0.34	0	100	0	0
20	0.342	0.34	0.002	100	0	0.2
40	0.344	0.34	0.004	100	0	0.4
60	0.346	0.34	0.006	100	0	0.6

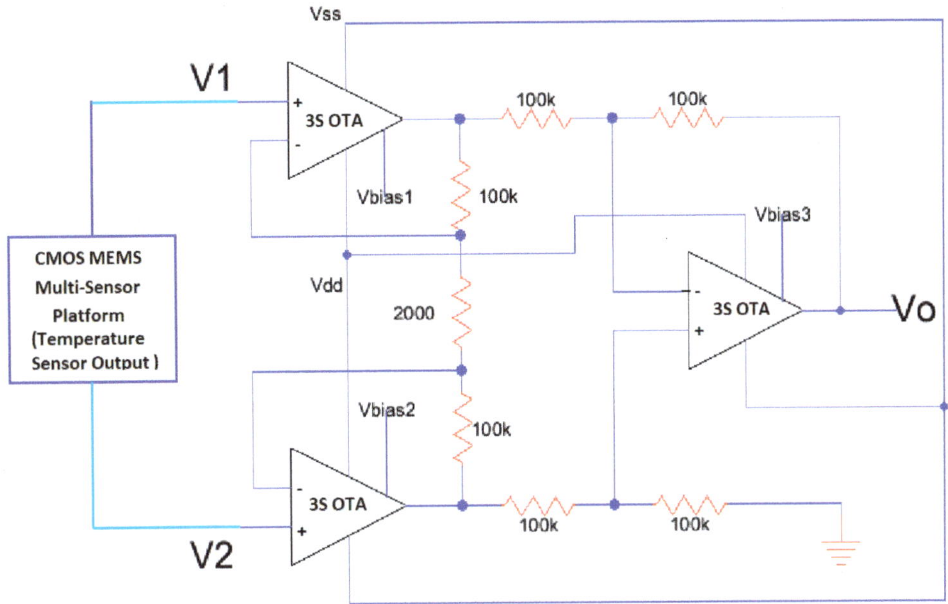

Figure 33. Instrumentation amplifier to condition the output from the temperature sensor described by the multiplatform of reference [42].

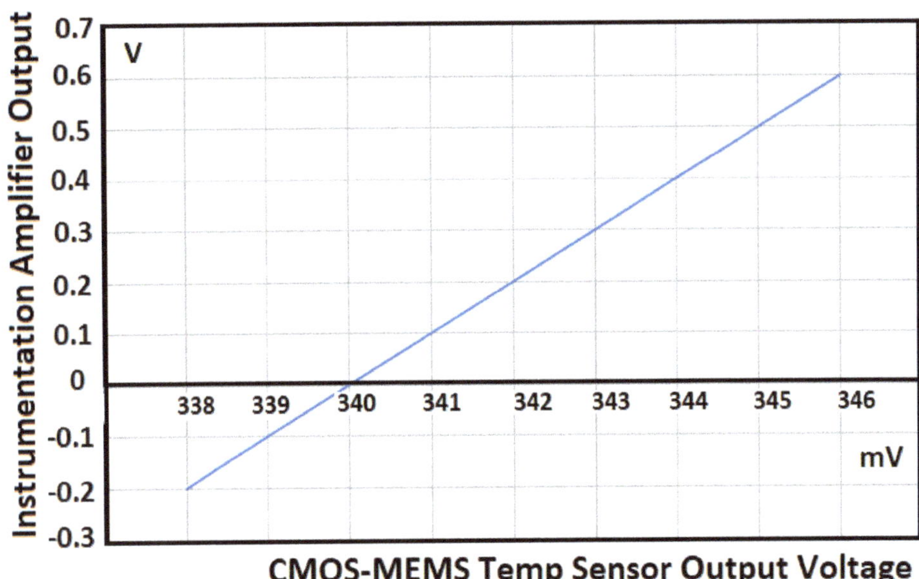

Figure 34. The linear signal conditioning performed by the instrumentation amplifier on signals coming from the CMOS-MEMS Temp Sensor Multiplatform [42].

The results obtained in this CMOS-MEMS sensor application confirm the possibility of using the EXCEL methods to perform first cut front end device design and testing. In teaching microelectronics, sometimes the use of high-performance software tools deviates attention toward the final objective and competence development for electronics engineers. The analysis of results from the previous two examples shows that the use of EXCEL methods provides a fertile background to start conceptual design processes without requiring complex or expensive software which is not readily available in many universities around the globe. Finally, to account for parasitic capacitances and temperature effects over the operating point of the amplifiers, Appendix A provides additional equations that can be added to the Excel method to find those fluctuations when using Spice simulations of the device. Appendix A.1 shows the parasitic capacitance modeling and Appendix A.2 illustrates the temperature model of this technology.

7. Conclusions

This paper described methods for developing the conceptual design of microelectronic circuits before performing schematic and layout simulations. The Excel methods developed here are useful for following guidelines and to make design decisions during iterative processes of designing where the number of specification requirements is larger than the degrees of freedom available to maneuver the design. The "thinking model" for paper and pencil calculations is described with the major roles and variants in the computation of electrical variables and sizes of the transistors that integrate analog microelectronic devices.

The Excel method may include straight single dimension, tabular, and two-dimensional methods that allow the design engineer to go step by step in the decision flow which navigates downwards in the Excel spreadsheet. Several examples of conceptual designs are shown with the corresponding spreadsheet diagram, circuit schematic, and layout design to perform simulation tests. The examples include passive components: resistors and capacitors, functional subcircuits such as primitive amplifiers, complete specialized amplifiers such as OTA devices, and term projects where students research specialized devices for biomedical instrumentation applications. The full-blown methodology includes the Excel method, schematic development, layout implementation, and simulation test, and it is used for conceptual design development in microelectronics courses at undergraduate and graduate levels. The undergraduate courses include industrial partner participation to develop instrumentation systems specified by industrial needs [43]; as a result, these educational collaborations with other entities have been applied in our academic educational practices and/or by applying our new educational model TEC21 at the undergraduate level [44–47]. As mentioned earlier, the main contributions of this research are in teaching CMOS microelectronics in which the method helps the instruction of the following points:

(1) Conceptual design evaluations using the traditional equations and preparation of layout implementation by setting up the schematic of the design.
(2) Providing possible design solutions, with layouts developed and comparing results with the thinking model specs to validate the third phase of the microelectronics design process.
(3) The methodology to develop complicated CMOS analog integrated circuits (IC) conceptual designs, specifically for undergraduate and graduate students.
(4) Development of complex conceptual designs using simple, non-expensive, and readily available software.
(5) Comparing a designed device with others published in the technical literature and used in biomedical instrumentation systems, particularly how far students are from major fundamental specifications and requirements such as gain, swing, frequency response, noise, and low power characteristics.
(6) Applying the designed amplifiers to CMOS-MEMS experimental devices in linear signal conditioning of multiplatform such as compound structures including Pirani, temperature, and gas sensors [42].

Finally, with the approach to teaching CMOS microelectronics and the use of readily available software such as Excel, it is possible to align teaching to reach the search for open innovation.

Author Contributions: G.D.-A., Conceptualization, methodology, simulation examples, validation, simulation validation, literature review, writing—original draft preparation, writing—review and editing, data processing, quantitative data collection, resources; J.M.R.-D., review conceptualization, methodology; O.I.G.P., Literature search, validation of the experimental methodology, review of data validation and simulations, analysis, writing, and editing. All authors have read and agreed to the published version of the manuscript.

Funding: This research received no external funding.

Institutional Review Board Statement: Not applicable.

Informed Consent Statement: Not applicable.

Acknowledgments: We wish to thank Steven Rubin and the crew that developed Electric-VLSI for the availability to have many professors, instructors and students working with this software in making conceptual designs of Microelectronic devices. Also, we want to acknowledge Analog Devices for the availability of its premier LTSpice software to professors, students and practicing engineers. Both software packages were very important in the validation of the conceptual designs developed by the Excel methods. Thanks must go to J. R. Baker and Philip Allen for their advice and workshops taken from 2005 until recently. Furthermore, we want to thank the Tecnológico de Monterrey, particularly the School of Engineering and Sciences (EIC), the Mechatronics and Electrical Engineering Department and the Computer Science Department for their faculty and student participation in the Microelectronic course TE-3034. In addition, a special thanks to the students who have participated in both courses, undergraduate and graduate level, since 2005 at Tecnológico de Monterrey, specially to Humberto González and Andrea Rodríguez who shared their term project design which was presented in this paper as a good example of using Excel methods to design microelectronic devices. Their participation enhanced the ideas and teaching strategies followed through the years while improving and refining the materials, simulation experiments and the fabless projects developed. Finally, the authors would like to acknowledge the Writing Lab, Institute for the Future of Education, Tecnologico de Monterrey, Mexico, for the financial and logistical support in the open access of this publication. This study belongs to the initiative R4C-IRG Reasoning for Complexity Interdisciplinary Research Group https://tec.mx/en/r4c-irg, accessed on 3 October 2021.

Conflicts of Interest: The authors declare no conflict of interest.

Appendix A

Appendix A.1. Parasitic Capacitances Modeling

In the analysis of parasitics inside the CMOS transistor, we can identify three types of capacitances. First, the thin oxide capacitance between the gate and the other terminals of the transistor that depends on the operating zone with an approximate value of

$$C_g = WLC_{ox} \tag{A1}$$

In the accumulation zone and factionary values of Cg of depletion, saturation, and linear zones of operation. The second type of parasitic capacitance inside the CMOS transistor is the PN junctions capacitors formed between the material P and N in the channel and substrate and sidewall capacitances. This is given by,

$$C_j \cong \frac{C_{jb}(A_b + A_{sw})}{\left(1 - \frac{v_j}{\varphi_B}\right)} \tag{A2}$$

The third type of parasitic capacitance inside a CMOS transistor is the Overlap capacitance that can be from the overlap of gates due to lateral diffusion (Cov) and fringing effects (Cf).

$$C_{ov} = C_{ox} \times L_D \tag{A3}$$

$$C_f = \frac{2\varepsilon_{ox}}{\pi} \ln\left(1 + \frac{T_{poly}}{t_{ox}}\right) \tag{A4}$$

$$C_{ol} = C_{ov} + C_f \tag{A5}$$

Table A1 illustrates the summary of the parasitic capacitances in a MOSFET at different operating zones. Moreover, Figure A1 depicts graphically the parasitic capacitance model of the CMOS devices.

Table A1. Parasitic MOSFET capacitance at different operation zones.

C/Zone	Cutoff	Linear	Saturation
C_{gs}	C_{ol}	$C_{ol} + 0.5C_g$	$C_{ol} + 0.67C_g$
C_{gd}	C_{ol}	$C_{ol} + 0.5C_g$	C_{ol}
C_{gb}	$[C_g\,C_{jc}\,]/[C_g + C_{jc}\,]$	0	0
C_{sb}	C_{jsb}	C_{jsb}	C_{jsb}
C_{db}	C_{jdb}	C_{jdb}	C_{jdb}
Subscripts: sb = source bulk	gb = gate bulk	db = drain bulk	ol = overlapp

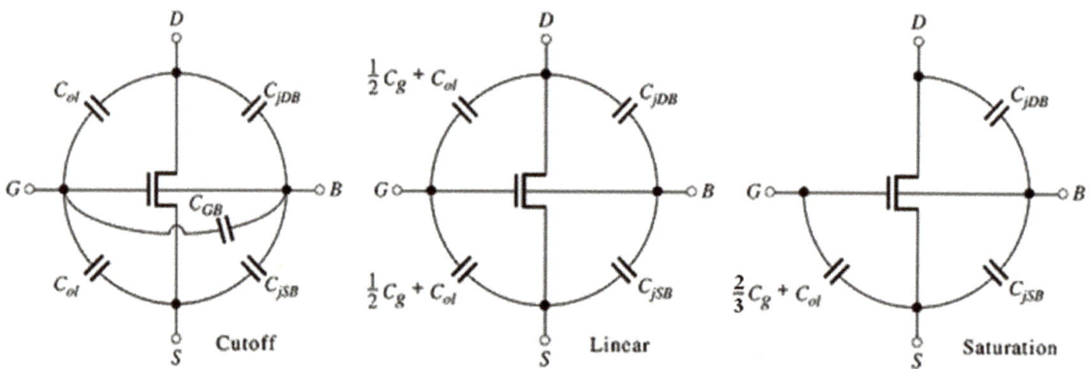

Figure A1. The parasitic capacitance model of the CMOS transistors.

Appendix A.2. Temperature Model and the Influence of Temperature in the Conceptual Design

In CMOS technology the influence of temperature in device conceptual designs is reflected mainly in the transconductance parameter and the threshold voltage, which can be evaluated using the Excel methods and validated using the ".temp" command in SPICE. The transconductance parameter changes with temperature as follows:

$$k(T) = k(T_0)[T/T_0]^{-1.5} \tag{A6}$$

The threshold voltage changes with temperature as follows:

$$V_T(T) \cong V_T(T_0) + \propto (T - T_0) \tag{A7}$$

Typically for NMOS transistors, $\propto_{NMOS}$ varies from -2 mV/°C to -3 mV/°C from 200 °K to 400 °K and for PMOS transistors the sign is reversed as one may expect. Therefore, the overall evaluations and calculations performed using the Excel method at room temperature would need to be re-calculated to account for the temperature range of operation. This can be validated using SPICE simulation as mentioned above.

For example, assume that we want to determine how the drain-source operating current of a CMOS 0.500 μm NMOS transistor varies from 27 °C to 100 °C. The device has $W/L = 5\,\mu/1\,\mu$, $k(T_0) = 117\,\mu A/V^2$, $V_{GS} = 2$ V, $V_T(T_0) = 0.8$ and $T_0 = 27\,°C$. The following calculations are performed:

(a)   At room temperature:

$$I_{DS}(27\,^\circ\text{C}) = \frac{117\mu\text{A/V}^2 \cdot 5\mu}{2\cdot 1\mu}(2 - 0.8)^2 = 421.2\ \mu\text{A}$$

(b)   At 100 °C (373 °K):

$$k(100\,^\circ\text{C}) = 117\mu\text{A}/V^2(373/300)^{-1.5} = 84.39\ \mu\text{A/V}^2$$

$$V_T(100\,^\circ\text{C}) = 0.8 - 0.002(73^\circ\text{C}) = 0.654\ \text{V}$$

$$I_{DS}(100\,^\circ\text{C}) = \frac{84.39\mu\text{A/V}^2 \cdot 5\mu}{2\cdot 1\mu}(2 - 0.654)^2 = 382.23\ \mu\text{A}$$

The previous example illustrates that a change of +73 °C in temperature produces a drop of 9.3% in the operating drain-source current of this device. Those calculations can be developed in the Excel methods to find the influence of temperature in the amplifier design. Those changes can also be verified by SPICE simulation using the temperature sweep accordingly.

References

1.   Rupp, K. 40 Years of Microprocessor Trend Data. Technische Universität Wien, Vienna Austria. Available online: https://www.karlrupp.net/2015/06/40-years (accessed on 10 May 2021).
2.   IBM Unveils World's First 2 Nanometer Chip Technology, Opening a New Frontier for Semiconductors. New York, NY. Available online: https://newsroom.ibm.com/2021-05-06-IBM-Unveils-Worlds-First-2-Nanometer-Chip-Technology,-Opening-a-New-Frontier-for-Semiconductors (accessed on 10 May 2021).
3.   Allen, P.; Holberg, J. *CMOS Analog Circuit Design*, 3rd ed. Oxford University Press: Oxford, UK, 2011.
4.   Baker, R.J. *CMOS Circuit Design, Layout and Simulation*, 3rd ed.; John Wiley & Sons: Hoboken, NJ, USA, 2019.
5.   Verbeeck, J.; Van Uffelen, M.; Steyaert, M.; Leroux, P. Conceptual Design of a MGy tolerant Integrated Signal Conditioning in 130 nm and 700 nm CMOS. *J. Instrum.* **2012**, *7*, C01017. [CrossRef]
6.   Henderson, S.C.; Johnson, E.W.; Janulis, J.R.; Tougaw, P.D. Incorporating standard CMOS design Process methodologies into the QCA logic design process. *IEEE Trans. Nanotechnol.* **2004**, *3*, 2–9. [CrossRef]
7.   Lee, T. *The Design of CMOS Radio Frequency Integrated Circuits*, 2nd ed.; Cambridge University Press: Cambridge, UK, 2004.
8.   Canovas-Carrasco, S.; Garcia-Sanchez, A.-J.; Garcia-Sanchez, F.; Garcia-Haro, J. Conceptual Design of a Nano-Networking Device. *Sensors* **2016**, *16*, 2104. [CrossRef] [PubMed]
9.   Grasso, A.D.; Marano, D.; Palumbo, G.; Pennisi, S. Design Methodology of Subthreshold Three-Stage CMOS OTAs Suitable for Ultra-Low-Power Low-Area and High Driving Capability. *IEEE Trans. Circ. Syst. I Regul. Pap.* **2015**, *62*, 1453–1462. [CrossRef]
10.   Aghighi, A.; Atkinson, J.; Bybee, N.; Anderson, S.; Crane, M.; Bailey, A.; Morell, R.; Hassanin, A.; Tajalli, A. CMOS Amplifier Design Based on Extended g_m/I_D Methodology. In Proceedings of the 2019 17th IEEE International New Circuits and Systems Conference (NEWCAS), Munich, Germany, 23–26 June 2019; pp. 1–4. [CrossRef]
11.   Dongyan, H.; Ming, Z.; Wei, Z. Design methodology of CMOS low power. In Proceedings of the 2005 IEEE International Conference on Industrial Technology, Hong Kong, China, 14–17 December 2005; pp. 114–118. [CrossRef]
12.   Green, M.M.; Pisani, M.B.; Dehollain, C. Design methodology for CMOS distributed amplifiers. In Proceedings of the 2008 IEEE International Symposium on Circuits and Systems, Seattle, WA, USA, 18–21 May 2008; pp. 728–731. [CrossRef]
13.   Hernández Sanabria, E.; Amaya Palacio, J.; Herrera, H.H.; Van Noije, W. A design methodology for an integrated CMOS instrumentation amplifier for bioespectroscopy applications. In Proceedings of the 2017 CHILEAN Conference on Electrical, Electronics Engineering, Information and Communication Technologies (CHILECON), Pucón, Chile, 29–31 October 2017; pp. 1–7. [CrossRef]
14.   Mohammed, M.; Abugharbieh, K.; Abdelfattah, M.; Kawar, S. Design methodology for MOSFET-based voltage reference circuits implemented in 28 nm CMOS technology. *AEU—Int. J. Electron. Commun.* **2016**, *70*, 568–577. [CrossRef]
15.   ON-Semiconductors. *MOSIS Wafer Electrical Tests Run V37P, Technology SCN05, Feature Size 0.5 Microns*; Technical Report; University of California´s Information Sciences Institute (ISI), University of Southern California: Marina del Rey, CA, December 2016.
16.   Lewyn, L.L.; Ytterdal, T.; Wulff, C.; Martin, K. Analog Circuit Design in Nanoscale CMOS Technologies. *Proc. IEEE* **2009**, *97*, 1687–1714. [CrossRef]
17.   Sadeqi, A.; Rahmani, J.; Habibifar, S.; Ammar Khan, M.; Mudassir Munir, H. Design Method for Two-Stage CMOS Operational Amplifier Applying Load/Miller Capacitor Compensation. 2020, pp. 153–162. Available online: https://mpra.ub.uni-muenchen.de/id/eprint/102931 (accessed on 1 September 2021).

18. Cheah, S.L.Y.; Yuen-Ping, H.O.; Shiyu, L.I. How the effect of opportunity discovery on innovation outcome differs between DIY laboratories and public research institutes: The role of industry turbulence and knowledge generation in the case of Singapore. *Technol. Forecast. Soc. Chang.* **2020**, *160*, 120250. [CrossRef]

19. Pearce, J.M. Economic savings for scientific free and open source technology: A review. *HardwareX* **2020**, *8*, e00139. [CrossRef] [PubMed]

20. Vicente-Saez, R.; Gustafsson, R.; Van den Brande, L. The dawn of an open exploration era: Emergent principles and practices of open science and innovation of university research teams in a digital world. *Technol. Forecast. Soc. Chang.* **2020**, *156*, 120037. [CrossRef]

21. Romero-Rodríguez, J.-M.; Ramírez-Montoya, M.-S.; Aznar-Díaz, I.; Hinojo-Lucena, F.-J. Social appropriation of knowledge as a key factor for local development and open innovation: A systematic review. *J. Open Innov.: Technol. Market Complex.* **2020**, *6*, 44. [CrossRef]

22. Beck, S.; Bergenholtz, C.; Bogers, M.; Brasseur, T.-M.; Conradsen, M.L.; Di Marco, D.; Distel, A.P.; Dobusch, L.; Dörler, D.; Effert, A.; et al. The Open Innovation in Science research field: A collaborative conceptualisation approach. *Ind. Innov.* **2020**, 1–50. [CrossRef]

23. Roman, M.; Liu, J.; Nyberg, T. Advancing the open science movement through sustainable business model development. *Ind. High. Educ.* **2018**, *32*, 226–234. [CrossRef]

24. Vicente-Saez, R.; Martinez-Fuentes, C. Open Science now: A systematic literature review for an integrated definition. *J. Bus. Res.* **2018**, *88*, 428–436. [CrossRef]

25. Schlagwein, D.; Conboy, K.; Feller, J.; Leimeister, J.M.; Morgan, L. "Openness" with and without Information Technology: A framework and a brief history. *J. Inf. Technol.* **2017**, *32*, 297–305. [CrossRef]

26. Friesike, S.; Widenmayer, B.; Gassmann, O.; Schildhauer, T. Opening science: Towards an agenda of open science in academia and industry. *J. Technol. Transf.* **2015**, *40*, 581–601. [CrossRef]

27. Razavi, B. *Fundamentals of Microelectronics*, 2nd ed.; Wiley: Hoboken, NJ, USA, 2014.

28. Dieck-Assad, G.; Vázquez-Piñón, M. *Simulation Practices for Electronics and Microelectronics Engineering, Volume 2, Chapter 5: CMOS Analog Microelectronics*; Amazon eBooks: Monterrey, Mexico, 2013.

29. Gray, P.; Hurst, P.J.; Lewis, S.H.; Meyer, R. *Analog Integrated Circuits*, 5th ed.; John Wiley & Sons: Hoboken, NJ, USA, 2011.

30. Lehman, T.; Cassia, M. 1V Power Supply CMOS CascodeAmplifier. *IEEE J. Solid State Circ.* **2001**, *36*, 1082–1086. [CrossRef]

31. Lee, M.; Cho, S.; Lee, N.; Kim, J. New Radiation-Hardened Design of a CMOS Instrumentation Amplifier and its Tolerant Characteristic Analysis. *Electronics* **2020**, *9*, 388. [CrossRef]

32. Lim, A.; Tan, A.; Kong, Z.-H.; Ma, K. A Design Methodology and Analysis for Transformer-Based Class-E Power Amplifier. *Electronics* **2019**, *8*, 494. [CrossRef]

33. Binkley, D.M. Tradeoffs and Optimization in Analog CMOS Design. In Proceedings of the 2007 14th International Conference on Mixed Design of Integrated Circuits and Systems, Date of Conference, Ciechocinek, Poland, 21–23 June 2007; pp. 47–60. [CrossRef]

34. González-Gómez, H.J.; Rodríguez-Vega, A. Micropower Op-Amps, Final Report for the TE-3034 Microelectronics Course Spring 2021 semester at Tecnológico de Monterrey, Monterrey, México, June 2021, unpublished work.

35. Lopez-Martin, A.J.; Ramirez-Angulo, J.; Carvajal, R.G.; Acosta, L. Micropower high current-drive class AB CMOS current-feedback operational amplifier. *Int. J. Circ. Theory Appl.* **2010**, *39*, 893–903. [CrossRef]

36. Pandit, S.; Mandal, C.; Patra, A. *Nano-Scale CMOS Analog Circuits: Models and CAD Techniques for High-Level Design*, 1st ed.; CRC Press: Boca Raton, FL, USA, 2014.

37. Rani, B.; Dhanoa, J.K. Designing of MOS Transistors in Two Stage OTA with Miller Compensation. *Int. J. Adv. Res. Electr. Electron. Instrum. Eng.* **2017**, *6*, 11, 8631–8637. [CrossRef]

38. Sadeqi, A.; Rahmani, J.; Habibifar, S. Design Method for Two-Stage CMOS Operational Amplifier Applying Load/Miller Capacitor Compensation. *Comput. Res. Prog. Appl. Sci. Eng. CRPASE: Trans. Electr. Electron. Comput. Eng.* **2020**, *6*, 153–162.

39. Qu, H. CMOS MEMS fabrication technologies and devices. *Micromachine* **2016**, *7*, 14. [CrossRef] [PubMed]

40. Peschot, A.; Qian, C.; Liu, T.J.K. Nanoelectromechanical Switches for Low-Power Digital Computing. *Micromachines* **2015**, *6*, 1046–1065. [CrossRef]

41. Huang, J.-Q.; Li, B.; Chen, W. A CMOS MEMS Humidity Sensor Enhanced by a Capacitive Coupling Structure. *Micromachines* **2016**, *7*, 74. [CrossRef] [PubMed]

42. Wang, J.; Yu, J. Multifunctional Platform with CMOS-Compatible Tungsten Microhotplate for Pirani, Temperature, and Gas Sensor. *Micromachines* **2015**, *6*, 1597–1605. [CrossRef]

43. Dieck-Assad, G.; Ávila-Ortega, A.; González Peña, O.I. Comparing Competency Assessment in Electronics Engineering Education with and without Industry Training Partner by Challenge-Based Learning oriented to Sustainable Development Goals. *Sustainability* **2021**, *13*, 10721. [CrossRef]

44. Dieck-Assad, G.; González Peña, O.I.; Rodríguez-Delgado, J.M. Evaluation of Emergency First Response's Competency in Undergraduate College Students: Enhancing Sustainable Medical Education in the Community for Work Occupational Safety. *Int. J. Environ. Res. Public Health* **2021**, *18*, 7814. [CrossRef]

45. González-Peña, O.I.; Morán-Soto, G.; Rodríguez-Masegosa, R.; Rodríguez-Lara, B.M. Effects of a Thermal Inversion Experiment on STEM Students Learning and Application of Damped Harmonic Motion. *Sustainability* **2021**, *13*, 919. [CrossRef]

46. González-Peña, O.I.; Peña-Ortiz, M.O.; Morán-Soto, G. Is It a Good Idea for Chemistry and Sustainability Classes to Include Industry Visits as Learning Outside the Classroom? An Initial Perspective. *Sustainability* **2021**, *13*, 752. [CrossRef]
47. Parra Cordova, A.; Gonzalez Pena, O.I. Enhancing student engagement with a small-scale car that is motion-controlled through chemical kinetics and basic electronics. *J. Chem. Educ.* **2020**, *97*, 3707–3713. [CrossRef]

Article

Convolution Kernel Operations on a Two-Dimensional Spin Memristor Cross Array

Saike Zhu [1,2], Lidan Wang [1,2,*], Zhekang Dong [3] and Shukai Duan [1,2]

[1]   School of Electronic Information Engineering, Southwest University, Chongqing 400715, China; saikezhu@email.swu.edu.cn (S.Z.); duansk@swu.edu.cn (S.D.)
[2]   Chongqing Key Laboratory of Nonlinear Circuits and Intelligent Information Processing, Chongqing 400715, China
[3]   School of Electronic Information, Hangzhou Dianzi University, Hangzhou 310018, China; englishp@hdu.edu.cn
*   Correspondence: ldwang@swu.edu.cn

Received: 15 October 2020; Accepted: 29 October 2020; Published: 31 October 2020

Abstract: In recent years, convolution operations often consume a lot of time and energy in deep learning algorithms, and convolution is usually used to remove noise or extract the edges of an image. However, under data-intensive conditions, frequent operations of the above algorithms will cause a significant memory/communication burden to the computing system. This paper proposes a circuit based on spin memristor cross array to solve the problems mentioned above. First, a logic switch based on spin memristors is proposed, which realizes the control of the memristor cross array. Secondly, a new type of spin memristor cross array and peripheral circuits is proposed, which realizes the multiplication and addition operation in the convolution operation and significantly alleviates the computational memory bottleneck. At last, the color image filtering and edge extraction simulation are carried out. By calculating the peak signal-to-noise ratio (PSNR) and structural similarity (SSIM) of the image result, the processing effects of different operators are compared, and the correctness of the circuit is verified.

Keywords: spin memristor; mask operation; memristor switch; memristor crossbar; image processing

---

1. Introduction

In recent years, in emerging data-intensive applications such as deep convolutional neural networks (DCNN) [1,2], image preprocessing tends to consume a lot of time and energy. The convolution processing operation is a widely applicable and far-reaching algorithm in the field of image processing. When convolution operations on images involve multiplication and addition operations, there will have problems of relatively large overhead, high real-time requirement, strong concurrency and frequent data exchange between memory and processor. The existing processing architecture has insufficient memory bandwidth and processing performance, resulting in a "memory wall" effect. Because the technological development of CMOS technology has reached its limit [3–5], the focus has been shifted from seeking faster processors to alleviating "memory bottlenecks". Although multicore architecture [6] and near-data structure [7,8] have been tried to improve convolution computing performance, they have not jumped out of the storage and computing separation architecture under the Von Neumann system, and there are still problems such as low energy/area efficiency, expensive hardware cost [9–11], etc. Memristor has the characteristics of integration of storage and calculation, so it has a wide range of application prospects in the field of in-memory computing.

Since 1971, Professor Chua predicted the existence of the fourth basic circuit device—memristor [12], and scientists have been searching for the missing memristor for 37 years.

Until 2008, HP laboratory [13] achieved the physical entity components of the memristor. Since then, new memristor devices and memristor models have been published [14,15]. The size of the memristor has reached the nanometer size, with good performance, storage and calculation integration, low energy consumption, short time, and good real-time performance in switching states. The memristance is controlled by electrical signals and has good non-volatile characteristics. It is compatible with CMOS technology in the production process. Therefore, the memristor has wide application prospects in the fields of image compression and filtering [16], image recognition [17], perceptron network [18], logic gate circuit [19,20], pulse neural network [21,22], non-volatile RAM [23], neural network synapse [24–26], establishment of chaotic circuits [27,28], and reconfigurable analog circuits [29], etc.

The spin memristor is a memristor model based on the magnetic domain wall movement mechanism. The resistance value will change only when the real-time current density is greater than or equal to the current density threshold. Therefore, the spin memristor has the characteristics of strong anti-interference ability and multi-level stability. Compared with memristors based on ion transport, it has better controllability [30]. Under the action of external excitation, the read and write speed of the spin memristor model can reach the nanosecond level. At the same time, the spin memristor has ultrafast spin dynamics [30–32] characteristics and good linearity, the smallest size, and compatibility with CMOS technology [7], so the spin memristor arrays are conducive to large-scale memory and high-speed logic operations. The spin memristor is a memristor model with a current threshold, so it has a more significant switching characteristic. The memristance value can quickly switch between high resistance and low resistance, and the constant voltage effect can be derived by mathematical deduction. The specific expression of the time required for the resistance change of the spin memristor under constant voltage can be obtained by mathematical deduction. Thus, the application time of external excitation can be effectively controlled and the unnecessary energy loss in memristive circuit can be reduced. Therefore, it can be designed as a flexible circuit switch.

Moreover, [33] proposed a memristor cross array, which greatly reduces the power consumption of the cross array and reduces the area of the cross array, but no specific application is given. A self-renewing mask circuit is proposed in [34], which uses a multibit memristor cross array to achieve convolution operation and realizes mean filtering and edge extraction. However, a 3×3 convolution kernel is used to perform sliding convolution on the entire image. During operation, each time the 3×3 image block is multiplied and added, the entire circuit needs to be used, which is very expensive. Moreover, when [35] used a 12×12 cross array, the 2D matrix is reduced to a 1D column vector, and multiple 2D convolution kernels are allowed to be read in parallel. The convolution operation of the convolution kernel is realized, but the peripheral circuit and control module are not designed.

This paper first proposes a logic switch circuit (MS) based on spin memristors, which realizes the AND gate operation to control the memristors cross array. By controlling the logic switch circuit, different numbers of convolution operators can be stored in the memristors cross array in the form of conductance. Since the cost of convolution operation mainly lies in the calculation of multiplication and addition, the pixels of the three channels of R, G, and B of the color image can be processed and converted into voltages, which input from the bottom to the memristor cross array to realize multiplication and addition operations, and realize different convolution operations. The memory bottleneck of computing is alleviated, and real-time parallel processing is realized. This paper verifies the correctness of the circuit through two image processing simulations. In the first simulation, the circuit is used to achieve the denoising and sharpening of the color image. The filtering effects of five image denoising operators, SRMC operator, median filter operator, 3×3 Gaussian filter operator, 5×5 Gaussian filter operator, and image sharpening operator, are compared. The second simulation is an edge detector, which simulates the extraction results of five image edge extraction operators: Prewitt operator, Sobel operator, Kirsch operator, Robert operator, and Laplacian operator. Two new types of operators based on Prewitt operator and Sobel operator are proposed. The image edge extraction effect has been significancly improved. By calculating peak signal-to-noise ratio (PSNR) and structural similarity (SSIM), the filtering and edge extraction effects of the above operators are compared.

The rest of the work in the paper is arranged as follows: Section 2 introduces the spin memristor model and proposes a logic switch based on the spin memristor. In Section 3, we will introduce the spin memristor cross array and its peripheral circuits that implement the convolution operation. In Section 4, the MATLAB simulation results of the circuit (mentioned in Section 3) in the image processing application are given, and the performances are analyzed and discussed respectively by adopting different filter operator and edge extraction operator. Finally, Section 5 summarizes the research content of this paper.

2. Logic Switch Based on Spin Memristor

The convolutional cross-array circuit proposed in this paper includes three significant modules—memristor cross-array, row/column address selector containing logic switches, and weighted average filter. The memristor cross-aray can store information in the form of conductance. By controlling the row/column address selector, when a voltage is applied to the corresponding column of the memristor cross-array, the current will carry the memristor information, and added together by an operational amplifier. The average filter can filter out polluted pixels in advance. This implementation has the advantages of parallelism, high efficiency, low complexity, fast speed, and easy control (low control voltage). This section will introduce the spin memristor model and the specific principles of spin memristor-based logic switches.

2.1. Introduction to Spin Memristors

The spin memristor is a device controlled by electric charge. Its structure diagram and equivalent circuit diagram are shown in Figure 1. The spin memristor is composed of a long spin-valve bar, which contains two layers of ferromagnets: The upper layer is the free layer, and the lower layer is the reference layer. The magnetic polarity of the reference layer is fixed in the magnetic layer by coupling technology. The free layer is divided into two by the magnetic domain wall. The two sections have opposite polarities. The resistance of each section is determined by the relative magnetization direction of the reference layer and free layer. When the magnetization directions of the two layers are opposite (parallel), the resistance of the spin memristor reaches the maximum (minimum).

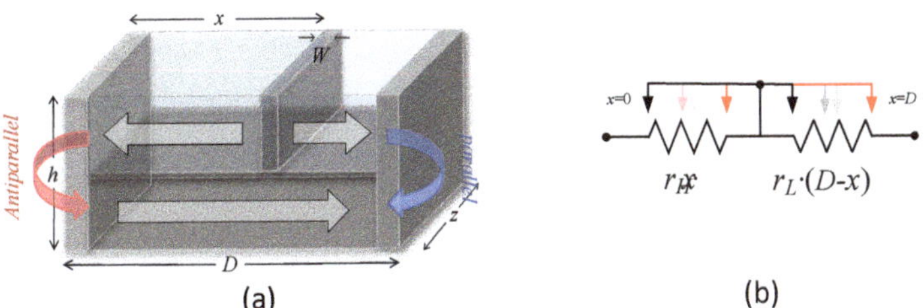

Figure 1. Spin memristor. (**a**) Structure diagram. (**b**) Equivalent circuit diagram.

The length, width, and height of the memristor in Figure 1a are represented by D, h, and z, respectively. W is the thickness of the magnetic domain wall. When power is applied to both ends of the spin memristor, the magnetic domain wall will move in the free layer, the total length of the two magnetization directions of the layer will change, leading to the change of the memristance value. Here, r_L and r_H respectively represent the high and low resistance values of the spin memristor, x represents the distance moved by the magnetic domain wall. Regardless of the domain wall width W, the equivalent circuit diagram of the memristor can be approximated by Figure 1b, and the resistance value of the memristor [15] is:

$$M(x) = r_H \cdot x + r_L(D - x) \tag{1}$$

The relationship between the moving speed v of the domain wall and the current density J is expressed as [15]:

$$v = \frac{dx}{dt} = \begin{cases} \Gamma_v \cdot J = \frac{\Gamma_v}{h_s \cdot z_s} \cdot \frac{dq}{dt}, & J \geq J_{cr} \\ 0, & J < J_{cr} \end{cases} \tag{2}$$

The relationship between x and the amount of charge passing through the memristor is:

$$x(t) = x_0 + \frac{\Gamma_v}{h \cdot z} \cdot q(t) \tag{3}$$

Equation (1) shows that the spintronic memristor has good linearity and can meet the requirements of multibit data storage. Here, Γ_v is the proportional coefficient, which is related to the structure of the device and the properties of the material. Besides, the adjustment of the memristance value $M_{(x)}$ is limited by the current density threshold. In other words, when the current density J is less than the critical current density J_{cr}, the resistance of the spin memristor is a constant. As long as the current density J is less than the critical current density, the state of the spintronic memristor will remain unchanged no matter how long the sensing voltage is maintained. When the current density J is higher than J_{cr}, the memristance value begins to change. The equation for calculating the current threshold J_{cr} of the device is as follows [15]:

$$J_{cr} = \frac{\alpha \gamma H_p e M_s}{P u_B} \sqrt{\frac{2A}{M_s H_k}} \tag{4}$$

where P represents the magnetic susceptibility of the material, M_s represents the saturation magnetization, α and γ represent the damping parameter and gyromagnetic ratio of the memristor, respectively. H_p and H_k represent the hard anisotropy and easy anisotropy of the magnetic material, respectively. A represents the exchange parameter, u_B is the Bohr magneton constant, e is the elementary charge. The expression of the current density J of the spin memristor device is [15]:

$$J = \frac{V}{M(x) \cdot h \cdot z} \tag{5}$$

As shown in Figure 2, the state variable x of the spin memristor model will gradually increase under the action of a positive voltage until the threshold condition is met or the maximum value of the state variable x is reached. The state variable x of the spin memristor model will gradually decrease under the action of negative voltage until the threshold condition is met or the minimum value of the state variable x is approached.

Figure 3 shows the resistance change curve of the spin memristor and its external excitation voltage curve. V is the voltage applied to the memristor, and $M_{(x)}$ represents the resistance value of the spin memristor when the magnetic domain wall moves by the distance x. Specifically, when the external excitation voltage is V_1 (red line), in the time domain [50, 100 ns], the external excitation voltage is positive and satisfied $V_1 / (r_H \cdot h \cdot z) \geq J_{cr}$, and the memristance value gradually increases to its maximum value; correspondingly, in the time domain [100 ns, 150 ns], when the external excitation voltage is negative and satisfied $|V_1 / (r_H \cdot h \cdot z)| \geq J_{cr}$, the memristance value gradually decreases to its minimum value. When the external excitation voltage is V_2 (blue line), in the time domain [50, 100 ns], the external excitation voltage is positive and satisfies the double inequality relationship $V_2 / (r_H \cdot h \cdot z) > J_{cr} > V_2 / (r_L \cdot h \cdot z)$, the memristance value gradually increases and the real-time current density decreases. If the real-time current density J is equal to the threshold current density J_{cr}, the memristor will remain unchanged. At this time, in the time domain [100 ns, 150 ns], only the polarity of the excitation voltage is changed, the resistance value of the memristor will not change. Appropriately increasing the amplitude of the excitation voltage to V_3 (green line), so that the external excitation voltage is negative and satisfied $|V_3/(r_H \cdot h \cdot z)| \geq J_{cr}$, the memristor resistance will decrease, and the real-time current density will increase and is always larger than the threshold current density J_{cr}, the memristance value eventually decreases to its minimum value. Similarly, the time required for the above-mentioned spin memristor resistance change can be derived from Equations (11) and (12).

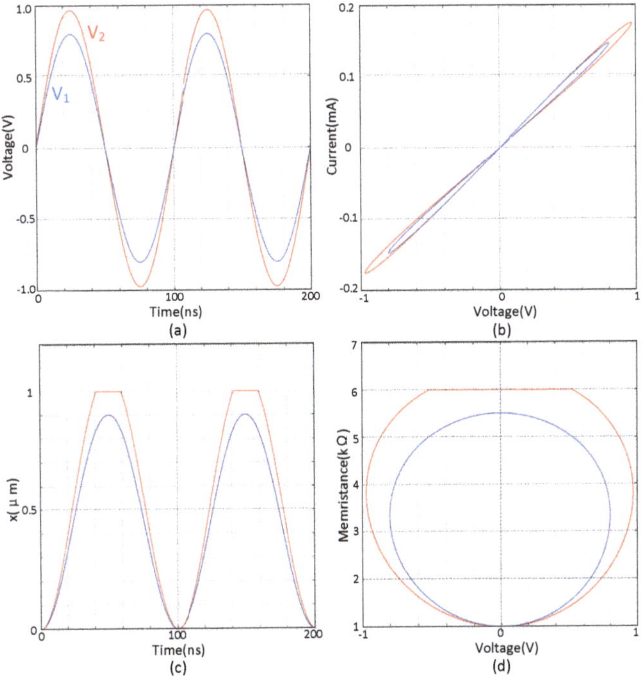

Figure 2. Simulation results of spin memristor ($V_1 < V_2$). (**a**) Time-Voltage curve, (**b**) Voltage-Current curve, (**c**) Time-Magnetic domain wall moving distance curve, (**d**) Voltage-Memristance curve.

The spin memristive damping parameter used in this paper is 0.002, and the current density threshold J_{cr} is approximately equal to 5.74×10^{12}, and the critical current can be calculated as $I_{cr} = J_{cr} \times h \times z = 4.018 \times 10^{-4}$A. After determining the structure and material of the device, the length D and width z of the spintronic memristor are the main factors for adjusting the high and low resistance of spin memristors. Besides, to reduce the localization of changes, the high-impedance resistance value of each cross-point memristor is set to 8 KΩ, and it is connected in series with a fixed resistance of 92 KΩ and a diode. The diode causes the current cross array to flow unidirectionally to prevent the sneak path current from affecting the output result. Adding a fixed resistor can prevent the memristance value from changing too small. In the following content, we will use "memristor" to refer to the memristor, resistor, and diode on the same contact, because the resistance part does not affect the training of the current control memristor. When a constant current exceeding the critical value is applied, the memristance value will decrease. Note that the adjustment in this study does not exceed one ns. By changing the critical current density J_{cr}, training current $I_{tr}(t)$, and speed coefficient v, the speed v of memristive decay can be adjusted.

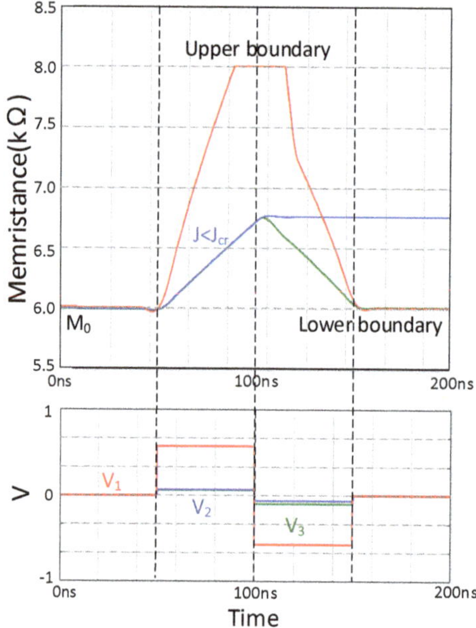

Figure 3. The change rule of the resistance value of the spin memristor model under the action of a voltage source of a fixed amplitude.

2.2. Memristor Switch (MS) Based on Magnetic Flux Control Spin Memristor

The resistance of the spin memristor can be described as [15],

$$M(x) = r_L \cdot D + (r_H - r_L) \cdot x \tag{6}$$

where M is the resistance value of the spin memristor, r_H and r_L are the high resistance and low resistance of the spin memristor, respectively, and D represents the length of the spin memristor. Spin memristor models all have unique threshold characteristics. That is, when the real-time current density J of the spin memristor is less than the threshold current density J_{cr}, it appears as a constant value resistance. On the contrary, when the real-time current density J of the spin memristor is greater than or equal to the threshold current density J_{cr}, the domain wall shifts and changes the resistance of the spin memristor. In this case ($J \geq J_{cr}$), perform differential operations on both sides of Equation (6), we can obtain:

$$\frac{dM}{dt} = (r_H - r_L) \cdot \Gamma_v \cdot J = \frac{(r_H - r_L) \cdot \Gamma_v}{h \cdot z} \cdot \frac{V}{M} \tag{7}$$

where Γ_v is the proportional coefficient, J represents the current density, h and z represent the width and height of the spin memristor, respectively, and V is the voltage applied on the spin memristor.

Then, integrate both sides of Equation (7) at the same time, to obtain the expression of the resistance value of the spin memristor controlled by the magnetic flux as follows:

$$M(\varphi) = \begin{cases} r_H, & \varphi > \varphi_{th2} \\ \sqrt{M_0^2 + 2B \cdot \varphi}, \varphi_{th1} \leq \varphi, & \varphi_{th1} \leq \varphi \leq \varphi_{th2} \\ r_L, & \varphi < \varphi_{th1} \end{cases} \tag{8}$$

where

$$\begin{cases} \varphi_{th1} = \frac{r_L^2 - M_0^2}{2B} \\ \varphi_{th2} = \frac{r_H^2 - M_0^2}{2B} \end{cases} \tag{9}$$

where M_0 is the initial resistance value of the spin memristor, φ is the magnetic flux flowing through the memristor, and its corresponding threshold φ_{th1}, φ_{th2} depends on the limit memristance value and the initial memristance value of the spin memristor. The auxiliary variable $B = \Delta r \cdot \Gamma_v / z \cdot h$ is a fixed constant.

In particular, based on Equation (8), the input voltage V is a constant with a fixed amplitude and the real-time current density of the memristor always satisfies $J \geq J_{cr}$. When the memristance changes from r_H to r_L, the total magnetic flux inside the spin memristor changes as follows:

$$\Delta\varphi = \frac{1}{2B}\left[r_H^2 - r_L^2\right] \tag{10}$$

The proposed AND logic switch (MS) based on the memristor is a simplified form of [36], which consists of two memristors P and Q connected through two positive terminals, as shown in Figure 4a. V_P and V_Q are two input voltages, and V_R is the output. In order to ensure the correctness of AND logic operation, $R_R \gg R_P, R_Q$. The truth table based on the AND operation of the memristor is shown in Figure 4b. The time required to change the memristance from r_L to r_H or from r_H to r_L is assumed to be the same. The initial states of the P and Q memristors are arbitrary. Based on $\Delta\varphi = V \cdot \Delta T$, the MS switching time T_1, the M_R switching time T_2 is

$$T_1 = \frac{1}{2B \cdot V_P}\left|r_H^2 - r_L^2\right| \tag{11}$$

$$T_2 = \frac{1}{2B \cdot V_R}\left|r_H'^2 - r_L'^2\right| \tag{12}$$

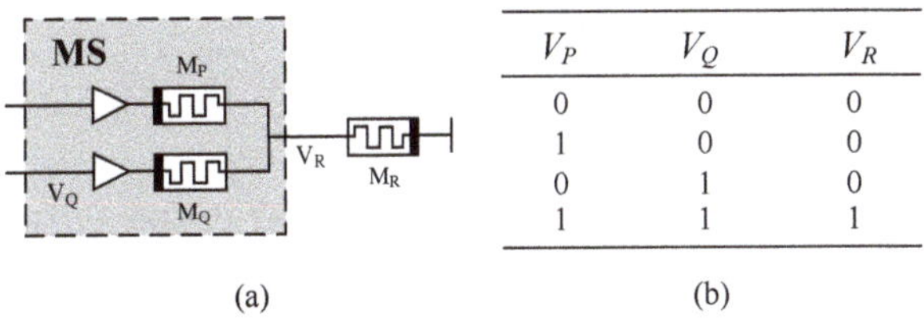

V_P	V_Q	V_R
0	0	0
1	0	0
0	1	0
1	1	1

(a) (b)

Figure 4. Logic switch based on spin memristor. (**a**) Logical switch based on memristor. (**b**) Truth table for AND operation.

r_H and r_L are the high resistance and low resistance of the memristor P and Q, respectively, and r_H' and r_L' are the high resistance and low resistance of the memristor M_R, as shown in Figure 4a. Assuming there is no threshold in MS. The output error V_e [36] is

$$V_e \approx \frac{R_Q}{R_P + R_Q}V_H = \frac{r_L}{r_H + r_L}V_H \tag{13}$$

According to the simulation results of the spin memristor logic switch in Figure 5, there are four special cases as follow:

1. $V_P = V_Q = V_H = $ "1" ("1" stands for logic 1, V_H stands for high-level voltage; "0" stands for logic 0, V_L stands for low-level voltage, and $V_L = 0$), the output voltage V_R is

$$V_R \approx \frac{R_Q + R_P}{R_P + R_Q} V_P = V_P \equiv V_H \tag{14}$$

Since there is a positive voltage across M_R, its memristive is reduced. After time T_2, $R_R = R_{on}$, the logic value stored in M_R changes to logic 1.

2. $V_P = V_H = "1"$, $V_Q = V_L = "0"$, the voltage across M_P is negative, and the voltage across M_Q is positive. After time T_1, $R_P = r_H, R_P = r_L (R_{on} \ll r_H)$, the output voltage V_R is

$$V_R \approx \frac{R_Q}{R_P + R_Q} V_P = \frac{r_L}{r_H + r_L} V_P \approx 0 \tag{15}$$

The logical value is stored in M_R to retain logic 0.

3. $V_P = V_L = "0"$, $V_Q = V_H = "1"$, the voltage across M_P is positive, and the voltage across M_Q is negative. After time T_1, $R_P = r_L$, $R_Q = r_H$, the output voltage V_R is

$$V_R \approx \frac{R_P}{R_P + R_Q} V_Q = \frac{r_L}{r_L + r_H} V_Q \approx 0 \tag{16}$$

The logical value is stored in M_R to retain logic 0.

4. $V_P = V_Q = V_L = "0"$. The output voltage $V_R = 0$, so the logic value is stored in M_R to retain logic 0. Therefore, the total time required for a complete logic switch operation is

$$T = T_1 + T_2 = \frac{1}{2B \cdot V_P} \left| r_H^2 - r_L^2 \right| + \frac{1}{2B \cdot V_R} \left| r_H'^{\,2} - r_L'^{\,2} \right| \tag{17}$$

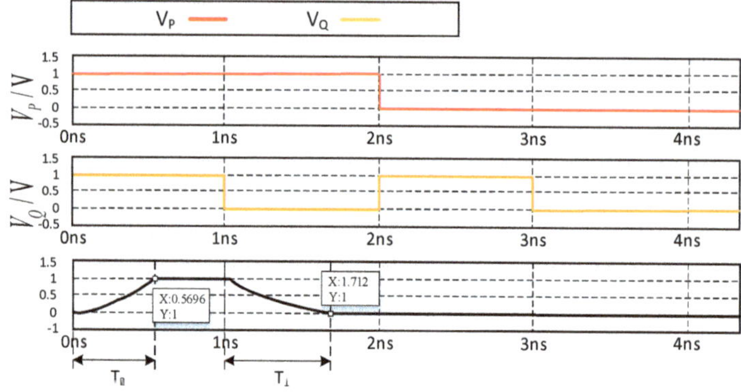

Figure 5. Simulation results of logic switches based on spin memristor.

3. Spin Memristor Cross-Array Circuit for Realizing Convolution Operation

The convolution kernel (also known as the filter) is usually a 2D matrix. To implement the convolution kernel on the memristor array, the 2D convolution kernel matrix is expanded and its dimensionality is reduced to 1D column vector. The paper uses the N × N convolution kernel matrix; here, we choose a convolution kernel with an odd number of N because if the filter size is even, the size of the input and output cannot be guaranteed unchanged. As shown in Figure 6, the convolution operation is performed on each pixel of the input image and involves three consecutive steps. When the convolution kernel is superimposed on the input image in this way, the operation starts. At first, the image to be processed is padded, and the center pixel of the convolution kernel is aligned with the single-pixel which is convolved in the input image. Then, multiply each pixel value in the input image by the corresponding value in the kernel. In the third step, the sum of the products of the second step is calculated, and the sum becomes the pixel value in the output. To convolve the entire image, it must be moved and received repeatedly to scan the input image pixel by pixel.

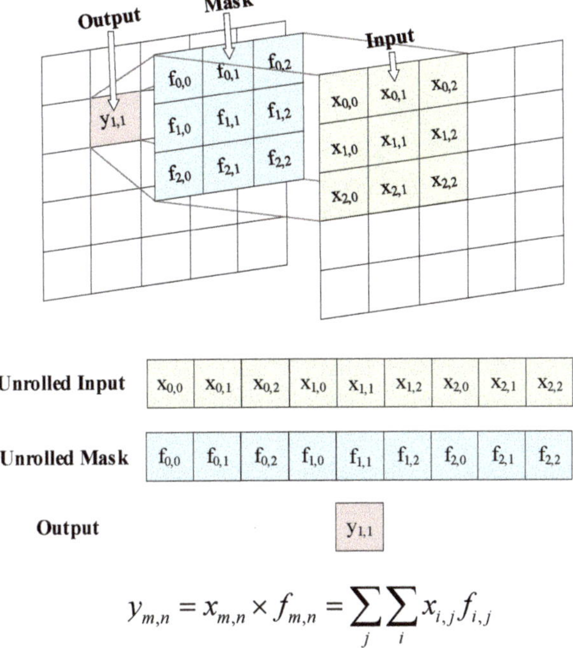

$$y_{m,n} = x_{m,n} \times f_{m,n} = \sum_{j}\sum_{i} x_{i,j} f_{i,j}$$

Figure 6. 2D matrix convolution.

3.1. Cross Array Circuit Based on Spin Memristor

Assume that a system is in discrete iteration V_k of the input, and the index is $V_k = 1, 2, \ldots, V_m$. During each iteration k, the system will receive a pair of two vectors M and N of size: the input image matrix $V_I^k \in R^M$ and the output image matrix $V_O^k \in R^N$. Assuming that the F convolution kernel array is an adjustable $N \times M$ matrix, and considering the estimator,

$$V_{out,j}^{(k)} = \sum_{i=1}^{M} F_{ji}^{(k)} V_{in,i}^{(k)} \tag{18}$$

where $i = 1, 2, \ldots, M$ and $j = 1, 2, \ldots, N$.

Figure 7 shows the memristor cross array used for convolution operations, which consists of a single crossbar array of $M_{(g,k)}$ and a $1/R_B$ constant term circuit. By expanding the pixels of an input image and converting them into voltages to be applied to rows, the currents read from multiple columns in an array can be calculated in parallel with the convolution results, which significantly accelerates the calculation speed. Here, $g_{i,j}$ is the memristor conductance at the intersection between the jth row and the kth column. $V_{in,j}$ is the input voltage applied to the jth column. $V_{C,K}$ is the row line voltage on the kth row. The row lines $V_{C,F}$ are connected to all the applied input voltages from $V_{in,1}$ to $V_{in,m}$ through R_B. In Figure 7, $V_{C,F}$, and negative feedback resistor R_1 enter G_F together. The latter constitutes an inverting amplifier. The output voltage of G_F is V_F, which is connected to all row lines from $V_{C,1}$ to $V_{C,n}$ through R_1. By applying Kirchhoff's current law to the row lines $V_{C,F}$, we can calculate V_F [33] as:

$$V_F = -\sum_{j=1}^{m} \frac{R_1}{R_B} V_{IN,j} \tag{19}$$

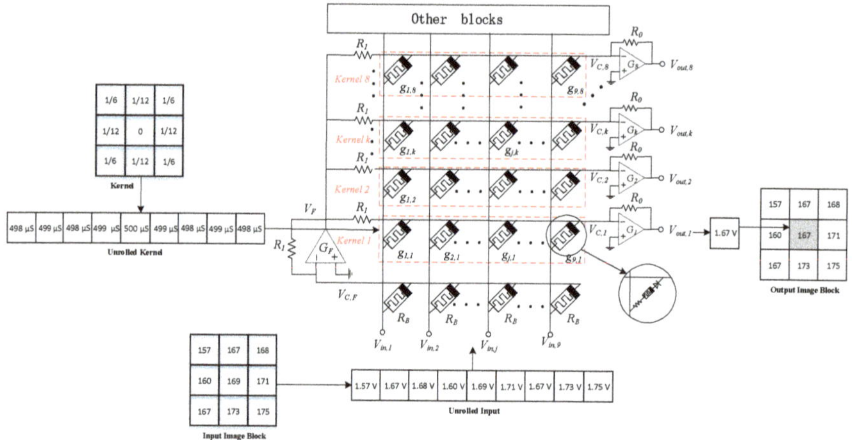

Figure 7. A memristor cross array for convolution operations; the red dotted boxes represent different convolution operators.

For the column lines, as shown in Figure 7, each row line is connected to its inverting amplifiers G_1 to G_n. For example, $V_{C,1}$ enters G_1 through negative feedback resistance R_0, $V_{out,1}$ is the output voltage of G_1, $V_{C,k}$ enters G_k similarly, and $V_{out,k}$ is the output voltage of G_k, $V_{out,k}$ can be calculated by the following Equation [33]:

$$V_{O,k} = -\left[\sum_{j=1}^{m}\left(R_0 \cdot g_{j,k} \cdot V_{IN,j}\right) + \frac{R_0}{R_1}V_F\right] \tag{20}$$

Incorporating Equation (19) into Equation (20), we can get:

$$
\begin{aligned}
V_{O,k} &= -\sum_{j=1}^{m}\left[R_0 \cdot g_{j,k} \cdot V_{IN,j} - \frac{R_0}{R_B}V_{IN,j}\right] \\
&= -\sum_{j=1}^{m} R_0 \cdot \left(\frac{1}{R_{j,k}} - \frac{1}{R_B}\right) \cdot V_{IN,j} \\
&= \sum_{j=1}^{m} R_0 \cdot \left(g_B - g_{j,k}\right) \cdot V_{IN,j}
\end{aligned}
\tag{21}
$$

If $\sum_{j=1}^{m} R_0 \cdot \left(g_B - g_{j,k}\right)$ is determined by the symmetric convolution kernel F in the kth row and the jth column, then we can rewrite Equation (21) as:

$$V_{O,k} = \sum_{j=1}^{m} F_{j,k} V_{IN,j} \tag{22}$$

$$F_{j,k} = R_0 \cdot \left(g_B - g_{j,k}\right) \tag{23}$$

Assuming that the conductance value range of the memristor is 0–1024 µS (where 0 means a minimal number, not 0 µS), the maximum number of digits represented by the memristor is 8 bit, so every 12 µS is equivalent to the number 1, assuming g_B is 500 µS and R_0 is 83 kΩ. As shown in Figure 7, a 3 × 3 input image block is randomly selected, and a 3 × 3 SRMC operator [2] is used for convolution operation here. First, expand the 3 × 3 SRMC operator into 1 × 9 rows, convert them into conductance values, and store them in the memristor cross-array. Then, expanding the input 3 × 3 image block pixel values into 1 × 9 rows, convert the 0–255 pixels into 0–2.55 V voltages, and input them into the memristor array through the column lines. The final V_{out}, one output voltage is 1.67 V, and the image pixel result calculated by convolution is 167, which verifies the correctness of the circuit.

3.2. Convolution Operation on Memristor Cross-Array Circuit

The state of the spin memristor is determined by the amount of charge or magnetic flux passing through it. Within the significant charge and magnetic flux range, if the total amount of charge or total magnetic flux passing through the spin memristor is zero, the memristor will eventually return to the initial value state [37].

Figure 8 shows the state change of the memristor whose initial state is x_0 caused by a symmetrical pulse current. The amplitude of the first half of the current is $-I_A$, and the width is T_W, which converts the spin memristor state to x_1 as follows:

$$x_1 = x_0 + \frac{\Gamma_v}{h \cdot z} \cdot (-I_A \cdot T_W) \tag{24}$$

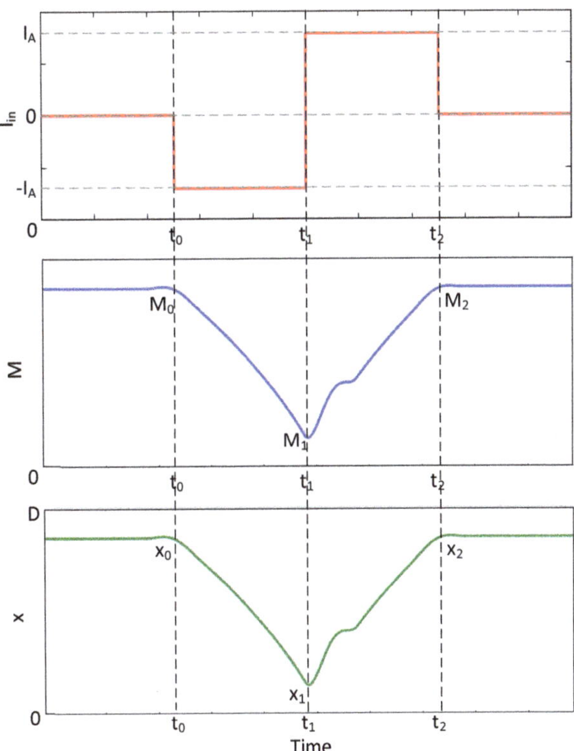

Figure 8. Symmetrical pulse current causes a change in the state of the memristor. From top to bottom are the input current I_{in}, the memristance value M, and the curve of the state variable x with time.

The amplitude of the second half current is I_A, and the width is T_W, and continue to switch the memristor state to x_2 as follows:

$$x_2 = x_1 + \frac{\Gamma_v}{h \cdot z} \cdot (I_A \cdot T_W) = x_0 \tag{25}$$

It can be seen that if the total amount of charge passing through the memristor is zero, the state of the memristor will eventually return to the initial state. If the total magnetic flux passing through the memristor is zero, the same effect will be achieved. For simplicity, suppose the power supply is a symmetrical pulse voltage with amplitudes $-V_A$ and V_A, and the widths of the two parts are both T_W. Using the relationship between the memristance value and magnetic flux in Equation (8), within the effective magnetic flux range, there are:

$$g_1 = \frac{1}{\sqrt{\frac{1}{g_0^2} - 2B\left(-V_A T_W\right)}} \tag{26}$$

$$g_2 = \frac{1}{\sqrt{\frac{1}{g_1^2} - 2B\left(-V_A T_W\right)}} = \frac{1}{\sqrt{\frac{1}{g_0^2} - 2B\left(-V_A T_W\right) - 2B\left(-V_A T_W\right)}} = g_0 \tag{27}$$

This property can be used for the read operation of the memristive memory. With this symmetrical sourse (current or voltage), the stored memristance value can be accurately read, so that the read operation will not adversely affect the stored memristance value. It is known that the resistance value of the memristor depends on the polarity, size, and time length of the external power. Assuming that the external power is a voltage pulse (amplitude is V_W, width is t_p), the write operation can set the memristor from the initial conductance g_0 to g' as follows:

$$g' = \frac{1}{\sqrt{\frac{1}{g_0^2} - 2B\varphi}} = \frac{1}{\sqrt{\frac{1}{g_0^2} - 2BV_w t_p}} \tag{28}$$

Using the symmetrical read voltage mode, the stored memristance value $g_{out} = i_o/v_o = g'$ can be read, and then V_W is

$$V_W = \frac{1}{2Bt_p}\left(\frac{1}{g_0^2} - \left(\frac{v_o}{i_o}\right)^2\right) \tag{29}$$

As shown in Figure 9, the memristor cross array circuit that implements the convolution operation includes a spin memristor cross array, row/column address selector, and address encoder, read/write control, read circuit, weighted average filter.

In the writing mode, the color image is decomposed into three channels of R, G, and B, decomposed into multiple $N \times N$ image blocks, expanded into $1 \times N^2$ rows, and then converted into voltage input from below, range for [0 V, 2.55 V], store the convolution kernel operator into the memristor array by controlling the address selector. The salt and pepper noise in the image is removed by a weighted average filter. When constructing the filter, the switching strategy is adopted to improve the filtering effect. In fact, the mean filter is called an algorithm, which makes an image blurry. In order to solve this problem, this paper proposes a switching strategy to find pixels that may be contaminated, and the mean filter is only applicable to noisy pixels. Specifically, the algorithm using the switching strategy has two-pixel thresholds p_1 and p_2, which represent the upper threshold and the lower threshold, respectively. Set $[p_1, p_2]$ to [5, 250], that is, $[V_{tu}, V_{tl}]$ to [0.05 V, 2.50 V]. First, execute the read mode to determine whether the current pixel is a noise point. If it is determined to be a normal pixel, inputs a voltage signal. Suppose the pixel value exceeds the range of $[p_1, p_2]$ (that is, the output voltage $V_{O(t)}$ exceeds the voltage range $[V_{tu}, V_{tl}]$ in the read mode, where V_{tu} and V_{tl} are the upper and lower threshold voltages), the pixel is considered as a noise point that should be processed by the filter. In the image edge extraction application, the weighted average filter in the dashed box in Figure 9 is not needed, so this part is deleted, reflecting the flexibility of the circuit.

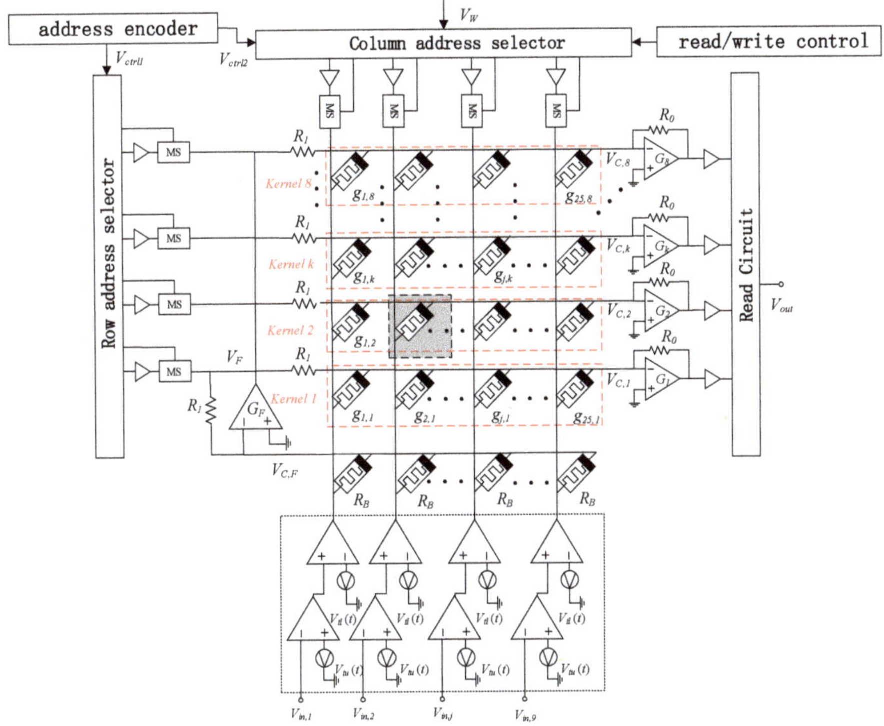

Figure 9. A memristor cross-array circuit for convolution operation. The dashed box is the weighted average filter.

Under the control of the address decoder and the row/column address selector, the input writes voltage pulse V_W is applied to the selected memristor to change its resistance state. When the write operation is over, the voltage across the memristor is zero, and the memristance value remains unchanged to realize the memory function.

In the read mode, when the read signal is valid, the read-write control circuit controls the selector to generate the symmetrical mode read voltage signal as described above and applies it to the target memristor. At this time, the read circuit can measure the flow through the memory resistor current and output it as an output signal. So far, the write and read operations of the memristor are completed.

To verify the effectiveness of the memristor simulation storage scheme proposed in this paper, computer simulations are carried out, and the memristor used is its mathematical model. The memristor parameters are set to $r_L = 5$ GΩ, $r_H = 6$ GΩ, $D = 1000$ nm, $h = 70$ nm, $z = 10$ nm, and J_{cr} is approximately equal to $5.74 \cdot e^{12}$. A write voltage pulse is applied to a certain memristor in the cross array, as shown in Figure 10a, and the corresponding memristance value change is shown in Figure 10c. The reading voltage as shown in Figure 10b is applied to the memristor, and the corresponding memristance value changes are shown in Figure 10d.

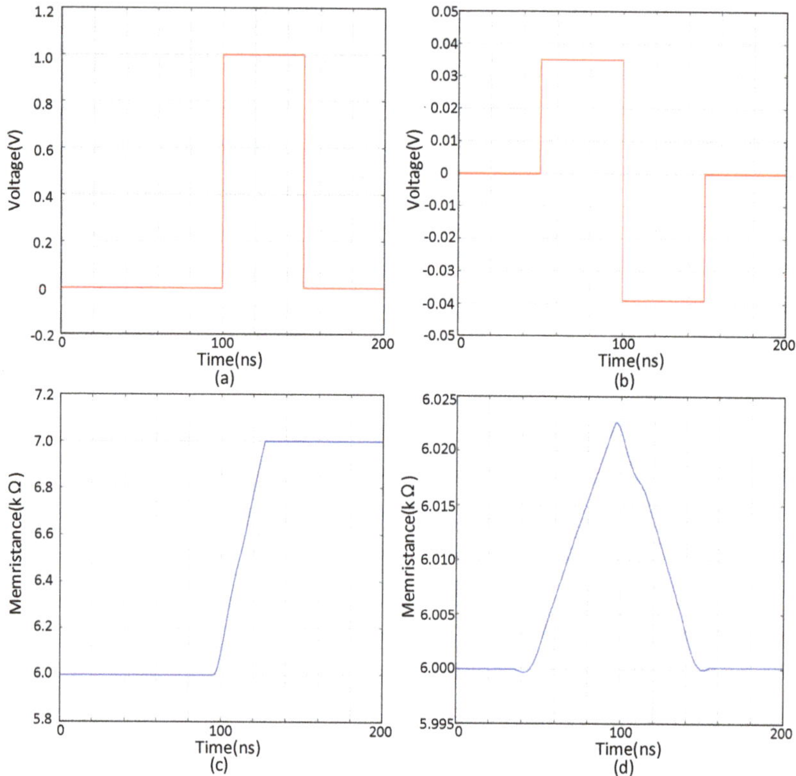

Figure 10. Write, read and restore operations of a charge-controlled spin memristor. (**a**) Write voltage pulse, (**b**) Read voltage pulse, (**c**) Write memristance value, (**d**) Read memristance value.

4. Application of Convolution Circuit in Color Image Denoising and Color Image Edge Extraction

By controlling the row/column address selector in Figure 9, 1–8 convolution operators can be stored in the memristor cross array in the form of conductance, and then the pixels of the image to be processed are converted into voltages. Voltages input from the bottom to the memristor cross array to realize multiplication and addition operations, which can realize different convolution operations, which reflects the flexibility of the circuit. This section introduces two image processing examples. In the first example, the circuit is used to achieve color image denoising and sharpening, and the filtering effects of different operators are compared. The second example is an edge detector, which uses a circuit structure that is faster and simpler than the previous circuit structure. This article uses MATLAB2020(a) version software for application-level simulation.

4.1. Color Image Denoising Based on Different Filter Operators

Figure 11 shows the convolution operator used in the weighted average filter, where the operator (a) [34] is implemented by two horizontal and vertical convolutions to store the convolution kernel operator in the memristor array in the training mode. Please note that the sum voltage is 1/12 times the read voltage, so the calculation is weighted, and no additional division is required. By continuously using the operator convolution twice, the memristor at the core of the mask can be modified to the expected state. First, assume that a memristor is a target. Set $[p_1, p_2]$ to [5, 250], that is, $[V_{tu}, V_{tl}]$ to [0.05 V, 2.50 V]. First, execute the read mode to determine whether the current pixel is a noise point. If it is determined to be a normal pixel, the value stored in the corresponding memristor will not change. Secondly, as described in the second section, sequentially store, train, and use convolution. Finally, by moving the image block, sliding processing is realized.

operator (a)

1/6	1/12	1/6
1/12	0	1/12
1/6	1/12	1/6

operator (b)

1/9	1/9	1/9
1/9	1/9	1/9
1/9	1/9	1/9

operator (c)

1/16	2/16	1/16
2/16	4/16	2/16
1/16	2/16	1/16

operator (d)

1/256	1/64	3/128	1/64	1/256
1/64	1/16	3/32	1/16	1/64
3/128	3/32	9/64	3/32	3/128
1/64	1/16	3/32	1/16	1/64
1/256	1/64	3/128	1/64	1/256

operator (e)

-1/256	-1/64	-3/128	-1/64	-1/256
-1/64	-1/16	-3/32	-1/16	-1/64
-3/128	-3/32	119/64	-3/32	-3/128
-1/64	-1/16	-3/32	-1/16	-1/64
-1/256	-1/64	-3/128	-1/64	-1/256

Figure 11. The convolution operator used in the weighted average filter. (**a**) SRMC operator [2]. (**b**) median filter operator. (**c**) 3×3 gaussian filter operator. (**d**) 5×5 gaussian filter operator. (**e**) image sharpening operator.

Operator (b) [38] is a median filter operator, taking the average of the nine values in the operator instead of the intermediate pixel value, so it has a smoothing and denoising effect. The gaussian filter operator presents a gaussian distribution in the horizontal and vertical directions, which highlights the weight of the center point after pixel smoothing has a better smoothing effect compared with mean filtering. Operator (c) is a 3×3 gaussian filter operator, operator (d) is a 5×5 gaussian filter operator [39], and operator (e) is an improved image sharpening operator based on the 5×5 gaussian filter operator. What is used is that the edge information in the image has higher contrast than the surrounding pixels, and this contrast is further enhanced after convolution, so that the image appears sharp and clear, which has the effect of sharpening the image.

Repeat these steps by applying the above operators until all pixels have removed the salt and pepper noise. We chose Lena as the primary material for the simulation. As shown in Figure 12a, 5% salt and pepper noise is added to the image in Figure 12b. Read the image that is denoised and stored in the original array in Figure 12b. Table 1 shows the power signal-to-noise ratio (PSNR) and structural similarity (SSIM) under different noise ratios. It means that the implementation has sufficient performance for different noise rates and gives correct results. Figure 12g is an improved gaussian filter operator, which achieves sharpened images. In short, the simulation results reported in this paper show that the circuit can implement a weighted average filter and give correct results.

Table 1. PSNR and SSIM for different operator.

Test Items	Operator(a)	Operator(b)	Operator(c)	Operator(d)	Operator(e)
PSNR(dB)	14.8294	15.8517	17.2997	16.8694	8.3064
SSIM	0.2569	0.4135	0.5943	0.5793	0.0481

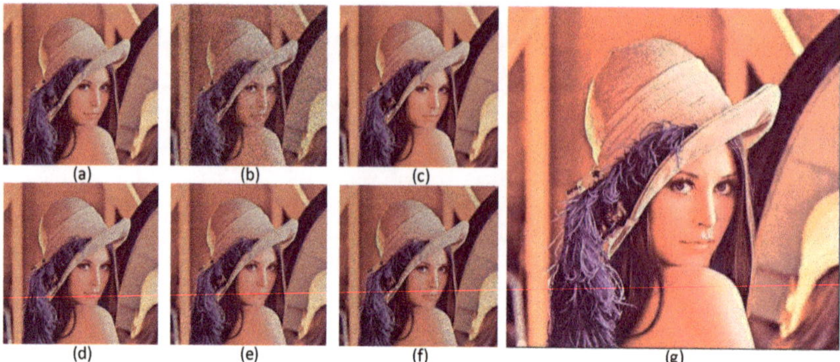

Figure 12. Filtered result. (**a**) Original image. (**b**) Lena added 5% salt and pepper noise. (**c**) Image denoised by SRMC operator. (**d**) Image denoised by median filter operator. (**e**) The image denoised by the 3×3 Gaussian filtering operator. (**f**) The image denoised by the 5×5 Gaussian filtering operator. (**g**) The image after the image sharpening operator.

4.2. Color Image Edge Detection Based on Different Convolution Operators

In the circuit of Figure 9, the weighted average filter is removed, and the input voltage directly enters the memristor array.

Figure 13 shows the operator used for color image edge extraction. Figure 13(a_1–a_4) is the prewitt operator [40], and Figure 9 (b_1–b_4) is the sobel operator [41], both of which are first-order differential operator. The former is an average filter, the latter is a weighted average filter, and the detected image edge may be more significant than 2 pixels. The advantage of these two methods is that the grayscale gradient and low-noise images have better detection effect, but the disadvantage is that the processing effect is not ideal for images with multiple complex noises. Based on this, the above two operators are improved, as shown in operators (a_1–a_8) and operators (b_1–b_8), diagonal operators in 45° and 135° directions are added on the basis of the original horizontal and vertical operators. The results are shown in Figure 14b–e, and the extracted edge details are more abundant than the original. The improved sobel operator works best and even retains some of the color details of the image.

Figure 13(c_1–c_8) is the kirsch operator [42]. The extraction result is shown in Figure 14f. Eight templates are used to convolve each pixel on the image to obtain the derivative. These eight templates represent eight directions, the maximum response to eight specific edge directions on the image, and the maximum value in the calculation (the weighted sum of 3×3 pixels is the sum after the corresponding position is multiplied), which is output as the edge of the image.

Figure 14 (d_1–d_8) is robert operator [43], also known as a cross differential operator. The advantage is that the positioning is accurate, and the operator is simple. The disadvantage is that to get a good effect, multiple operators are required to superpose, the calculation speed is slow, and it is more sensitive to noise. The extraction result is shown in Figure 14g, and the extraction effect is average.

Figure 14 (e_1–e_4) are laplacian operators. Laplacian operators are more sensitive to noise, so the image is generally smoothed first. Because smoothing is also performed using templates, the usual segmentation algorithms are based on laplacian operators and the smoothing operator, which combined to generate a new template. It can be seen from the extraction result of Figure 14g that in addition to the edge extraction effect, the template also has the effect of sharpening the picture.

Figure 13. Convolution operator for edge detection. The blue arrow indicates the direction of current, the number sign determines the polarity of the voltage. (a_1–a_4) Prewitt operator (in the yellow box). (a_1–a_8) Proposed Prewitt operator (in the green box). (b_1–b_4) Sobel operator (in the blue box). (b_1–b_8) Proposed Sobel operator (in the red box). (c_1–c_8) Kirsch operator. (d_1–d_8) Robert operator. (e_1–e_4) Laplacian operator.

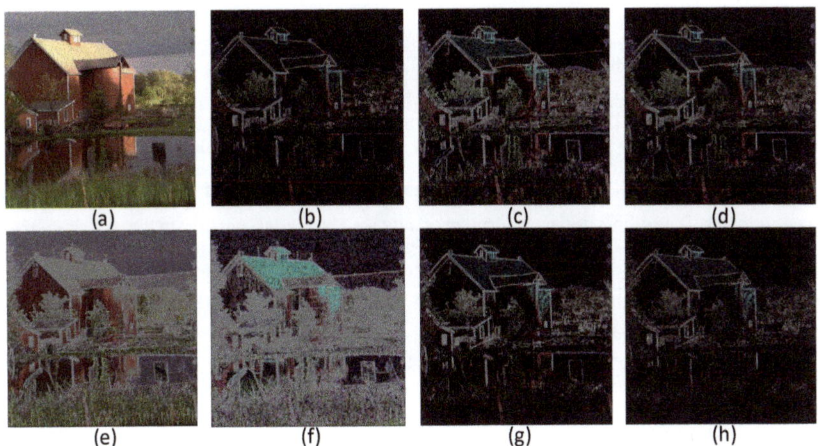

Figure 14. The realization of color image edge extraction.(a) Original image, (b) Prewitt operator, (c) Proposed Prewitt operator, (d) Sobel operator, (e) Proposed Sobel operator, (f) Kirsch operator, (g) Robert operator, (h) Laplacian operator.

Use the above operators to process the pixel values stored in the memristor array. In the case where the input circuit provides only one summation voltage at the same time, a negative number is included in the convolution. In order to avoid storing negative values (i.e., negative pixels) in the target memristor, we first use positive convolution to limit the negative result to 0 and maintain the positive result normally. In addition, the resulting pixel value has a risk of 255. This problem can be solved by limiting the maximum (current) output of VCCS at 13.209 µA By sliding the image block repeatedly, the edge of the color image can be extracted effectively until all pixels are processed by edge detector. We chose a 256 × 256 pixel picture as the primary material for this simulation. The result is shown in Figure 14. Table 2 shows the power signal-to-noise ratio (PSNR) of the edge extraction results after processing by different operators.

Table 2. PSNR for different operator.

Test Items	Prewitt	Proposed Prewitt	Soble	Proposed Soble	Kirsch	Robert	Laplacian
PSNR(dB)	7.7960	8.8183	8.2401	15.5461	11.2648	8.4652	8.0827

5. Conclusions

This paper proposes a generalized circuit scheme based on the filtering convolution operator and the edge extraction convolution operator to implement image processing applications on the spin memristor crossover to alleviate the storage bottleneck of data-intensive applications. By adopting a self-updating circuit and parallel multi-bit selective adder and convolution algorithm, we implement color image filtering and edge extraction with different operators, and reduce the dependence on data exchange to the lowest level. All the devices used in this paper are compatible with CMOS technology, so the proposed implementation scheme also shows the advantages of large-scale integrated manufacturing. The practicability and excellent performance of this work in image processing have been proved by algorithm simulation.

Author Contributions: Conceptualization, S.Z. and Z.D.; methodology, S.Z., Z.D. and L.W.; software, S.Z.; validation, S.Z., Z.D. and L.W.; formal analysis, S.Z. and L.W.; investigation, S.Z.; resources, L.W.; data curation, S.Z. and L.W.; writing–original draft preparation, S.Z.; writing–review and editing, S.Z., Z.D. and L.W.; visualization, S.Z.; supervision, L.W., Z.D. and S.D.; project administration, L.W. and S.D.; funding acquisition, L.W. and S.D. All authors have read and agreed to the published version of the manuscript.

Funding: This research was funded by the National Key R&D Program of China (Grant No. 2018YFB1306600), the National Natural Science Foundation of China (Grant Nos.62076207, 62076208), the Fundamental Science and Advanced Technology Research Foundation of Chongqing, China (Grant Nos. cstc2017jcyjBX0050).

Acknowledgments: The authors would like to thank the National Key R&D Program of China, the National Natural Science Foundation of China, the Fundamental Science and Advanced Technology Research Foundation of Chongqing, Chongqing Key Laboratory of Nonlinear Circuits and Intelligent Information Processing, School of Electronic Information Engineering, Southwest University, and all the partners (Lidan Wang, Shukai Duan, Zhekang Dong and Chunyan Ren) for their support in the conceptualization and revision of the manuscript.

Conflicts of Interest: The authors declare no conflict of interest. The funders support in the collection, analyses or interpretation of data; in the writing of the manuscript, or in the decision to publish the result

Abbreviations

The following abbreviations are used in this manuscript:

MDPI	Multidisciplinary Digital Publishing Institute
DOAJ	Directory of open access journals
TLA	Three letter acronym
LD	Linear dichroism

References

1. Chen, B.; Polatkan, G.; Sapiro, G.; Blei, D.; Dunson, D.; Carin, L. Deep learning with hierarchical convolutional factor analysis. *IEEE Trans. Pattern Anal. Mach. Intell.* **2013**, *35*, 1887–1901. [CrossRef]
2. Dong, C.; Loy, C.C.; He, K.; Tang, X. Image super-resolution using deep convolutional networks. *IEEE Trans. Pattern Anal. Mach. Intell.* **2016**, *38*, 295–307. [CrossRef] [PubMed]
3. Kaxiras, S. *Architecture at the End of Moore*; Springer: New York, NY, USA, 2012.

4. Esmaeilzadeh, H.; Blem, E.; Amant, R.S.; Sankaralingam, K.; Burger, D. Dark silicon and the end of multicore scaling. In Proceedings of the 2011 38th Annual international symposium on computer architecture (ISCA), San Jose, CA, USA, 4–8 June 2011; pp. 365–376.

5. Taur, Y. CMOS design near the limit of scaling. *IBM J. Res. Dev.* **2002**, *46*, 213–222. [CrossRef]

6. Sheikh, H.F.; Ahmad, I.; Fan, D. An evolutionary technique for performance-energy-temperature optimized scheduling of parallel tasks on multicore processors. *IEEE Trans. Parallel Distrib. Syst.* **2016**, *27*, 668–681. [CrossRef]

7. Farmahini-Farahani, A.; Ahn, J.H.; Morrow, K.; Kim, N.S. DRAMA: An architecture for accelerated processing near memory. *IEEE Comput. Archit. Lett.* **2015**, *14*, 26–29. [CrossRef]

8. Azarkhish, E.; Pfister, C.; Rossi, D.; Loi, I.; Benini, L. Logic-base interconnect design for near memory computing in the smart memory cube. *IEEE Trans. Very Large Scale Integr. (VLSI) Syst.* **2017**, *25*, 210–223. [CrossRef]

9. Xue, W.; Yang, C.; Fu, H.; Wang, X.; Xu, Y.; Gan, L.; Lu, Y.; Zhu, X. Enabling and scaling a global shallow-water atmospheric model on Tianhe-2. In Proceedings of the 2014 IEEE 28th International Parallel and Distributed Processing Symposium, Phoenix, AZ, USA, 19–23 May 2014; pp. 745–754.

10. Zhang, X.; Yang, C.; Liu, F.; Liu, Y.; Lu, Y. Optimizing and scaling HPCG on Tianhe-2: Early experience. In Proceedings of the 14th International Conference on Algorithms and Architectures for Parallel Processing (ICA3PP), Dalian, China, 24–27 August 2014; pp. 28–41.

11. Bell, G.; Gray, J. What's next in high-performance computing? *Commun. ACM* **2002**, *45*, 91–95. [CrossRef]

12. Chua, L.O. Memristor—The missing circuit element. *IEEE Trans. Circuit Theory* **1971**, *5*, 507–519. [CrossRef]

13. Strukov, D.B.; Snider, G.S.; Stewart, D.R.; Williams, R.S. The missing memristor found. *Nature* **2008**, *453*, 80–83. [CrossRef]

14. Kvatinsky, S.; Ramadan, M.; Friedman, E.G.; Kolodny, A.; Williams, R.S. VTEAM: A general model for voltage-controlled memristors. *IEEE Trans. Circuits Syst. II Express Briefs* **2015**, *62*, 786–790. [CrossRef]

15. Chen, Y.C.Y.; Wang, X.W.X. Compact modeling and corner analysis of spintronic memristor. In Proceedings of the 2009 IEEE/ACM International Symposium on Nanoscale Architectures, San Francisco, CA, USA, 30–31 July 2009.

16. Li, C.; Hu, M.; Li, Y.; Jiang, H.; Ge, N.; Montgomery, E.; Zhang, J.; Song, W.; Dávila, N.; Graves, C.E.; et al. Analogue signal and image processing with large memristor crossbars. *Nat. Electron.* **2017**, *1*, 52. [CrossRef]

17. Yao, P.; Wu, H.; Gao, B.; Tang, J.; Zhang, Q.; Zhang, W.; Yang, J.J.; Qian, H. Fully hardware-implemented memristor convolutional neural network. *Nature* **2020**, *577*, 641–646. [CrossRef]

18. Cai, F.; Correll, J.M.; Lee, S.H.; Lim, Y.; Bothra, V.; Zhang, Z.; Flynn, M.P.; Lu, W.D. A fully integrated reprogrammable memristor-CMOS system for efficient multiply-accumulate operations. *Nat. Electron.* **2019**, *2*, 290–299. [CrossRef]

19. Dong, Z.; Qi, D.; He, Y.; Xu, Z.; Hu, X.; Duan, S. Easily Cascaded Memristor-CMOS Hybrid Circuit for High-Efficiency Boolean Logic Implementation. *Int. J. Bifurc. Chaos* **2018**, *27*, 1850149. [CrossRef]

20. Dong, Z.; He, Y.; Hu, X.; Qi, D.; Duan, S. Flexible memristor-based LUC and its network integration for Boolean logic implementation. *IET Nanodielectr.* **2019**, *2*, 61–69. [CrossRef]

21. Dong, Z.; Lai, C.S.; Qi, D.; Xu, Z.; Li, C.; Duan, S. A general memristor-based pulse coupled neural network with variable linking coefficient for multi-focus image fusion. *Neurocomputing* **2018**, *308*, 172–183. [CrossRef]

22. Dong, Z.; Zhang, S.; Ma, B.; Qi, D.; Luo, L.; Zhou, M. A Hybrid Multi-Frame Super-Resolution Algorithm Using Multi-Channel Memristive Pulse Coupled Neural Network and Sparse Coding. In Proceedings of the 2019 7th International Conference on Information, Communication and Networks (ICICN), Macao, China, 24–26 April 2019.

23. Duan, S.; Hu, X.; Wang, L.; Li, C.; Mazumder, P. Memristor-based RRAM with applications. *Sci. China Inf. Sci.* **2012**, *55*, 1446–1460. [CrossRef]

24. Wang, L.; Li, H.; Duan, S.; Huang, T.; Wang, H. Pavlov associative memory in a memristive neural network and its circuit implementation. *Neurocomputing* **2016**, *171*, 23–29. [CrossRef]

25. Duan, S.; Wang, H.; Wang, L.; Huang, T.; Li, C. Impulsive Effects and Stability Analysis on Memristive Neural Networks With Variable Delays. *IEEE Trans. Neural Netw. Learn. Syst.* **2017**, *28*, 476–481. [CrossRef]

26. Duan, S.; Hu, X.; Dong, Z.; Wang, L.; Mazumder, P. Memristor-Based Cellular Nonlinear/Neural Network: Design, Analysis, and Applications. *IEEE Trans. Neural Netw. Learn. Syst.* **2015**, *26*, 1202–1213. [CrossRef]

27. Wang, L.; Drakakis, E.; Duan, S.; He, P.; Liao, X. Memristor Model and Its Application for Chaos Generation. *Int. J. Bifurc. Chaos* **2012**, *22*, 1250205. [CrossRef]

28. Wang, L.; Duan, S. A Chaotic Attractor in Delayed Memristive System. *Abstr. Appl. Anal.* **2012**, 2012, 726927. [CrossRef]

29. Pershin, Y.V.; Ventra, M.D. Practical approach to programmable analog circuits with memristors. *IEEE Trans. Circuits Syst. Regul. Pap.* **2010**, *57*, 1857–1864. [CrossRef]

30. Pershin, Y.V.; Ventra, M.D. Spin memristive systems: Spin memory effects in semiconductor spintronics. *Phys. Rev. B Condens. Matter* **2008**, *78*, 113309. [CrossRef]

31. Saidl, V.; Němec, P.; Wadley, P.; Hills, V.; Campion, R.P.; Novák, V.; Edmonds, K.W.; Maccherozzi, F.; Dhesi, S.S.; Gallagher, L.B.; et al. Optical determination of the Néel vector in a CuMnAs thin-film antiferromagnet. *Nat. Photonics* **2017**, *11*, 91–96. [CrossRef]

32. Zhao, W.; Ravelosona, D.; Klein, J.O.; Chappert, C. Domain Wall Shift Register-Based Reconfigurable Logic. *IEEE Trans. Magn.* **2011**, *47*, 2966–2969. [CrossRef]

33. Truong, S.N.; Min, K.S.T. New Memristor-Based Crossbar Array Architecture with 50-% Area Reduction and 48-% Power Saving for Matrix-Vector Multiplication of Analog Neuromorphic Computing. *J. Semicond. Technol. Sci.* **2014**, *14*, 356–363. [CrossRef]

34. Shang, L.; Duan, S.; Wang, L.; Huang, T. SRMC: A multibit memristor crossbar for self-renewing image mask. *IEEE Trans. Very Large Scale Integr. (VLSI) Syst.* **2018**, *26*, 2830–2841. [CrossRef]

35. Gao, L.; Chen, P.Y.; Yu, S. Demonstration of Convolution Kernel Operation on Resistive Cross-Point Array. *IEEE Electron. Device Lett.* **2016**, *37*, 870–873. [CrossRef]

36. Zhang, Y.; Shen, Y.; Wang, X.; Cao, L. A novel design for memristor based logic switch and crossbar circuits. *IEEE Trans. Circuits Syst. I Reg. Pap.* **2015**, *62*, 1402–1411. [CrossRef]

37. Ho, Y.; Huang, G.M.; Li, P. Dynamical Properties and Design Analysis for Nonvolatile Memristor Memories. *Circuits Syst. Regul. Pap. IEEE Trans.* **2011**, *58*, 724–736. [CrossRef]

38. Huang, T.; Yang, G.J.T.G.Y.; Tang, G. A fast two-dimensional median filtering algorithm. *IEEE Trans. Acoust. Speech Signal Process.* **1979**, *27*, 13–18. [CrossRef]

39. Haddad, R.A.; Akansu, A.N. A class of fast Gaussian binomial filters for speech and image processing. *IEEE Trans. Signal Process.* **1991**, *39*, 723–727. [CrossRef]

40. Prewitt, J.M. Object enhancement and extraction. *Pict. Process. Psychopictorics* **1970**, *10*, 15–19.

41. Farid, H.; Simoncelli, E.P. Optimally rotation-equivariant directional derivative kernels. In Proceedings of the International Conference on Computer Analysis of Images and Patterns, Kiel, Germany, 10–12 September 1997; pp. 207–214.

42. Kirsch, R.A. Computer determination of the constituent structure of biological images. *Comput. Biomed. Res.* **1971**, *4*, 315–328. [CrossRef]

43. Tippett, J.T.; Borkowitz, D.A.; Clapp, L.C.; Koester, C.J.; Vanderburgh, A., Jr. *Optical and Electro-Optical Information Processing*; Massachusetts Inst of Tech Cambridge: Cambridge, MA, USA, 1965.

Publisher's Note: MDPI stays neutral with regard to jurisdictional claims in published maps and institutional affiliations.

MDPI

St. Alban-Anlage 66

4052 Basel

Switzerland

Tel. +41 61 683 77 34

Fax +41 61 302 89 18

www.mdpi.com

Sensors Editorial Office

E-mail: sensors@mdpi.com

www.mdpi.com/journal/sensors